T0324732

A Chorus of Bells and
Other Scientific Inquiries

A Chorus of Bells and Other Scientific Inquiries

Jeremy Bernstein
Stevens Institute of Technology, USA

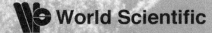

World Scientific

NEW JERSEY · LONDON · SINGAPORE · BEIJING · SHANGHAI · HONG KONG · TAIPEI · CHENNAI

Published by

World Scientific Publishing Co. Pte. Ltd.
5 Toh Tuck Link, Singapore 596224
USA office: 27 Warren Street, Suite 401-402, Hackensack, NJ 07601
UK office: 57 Shelton Street, Covent Garden, London WC2H 9HE

Library of Congress Cataloging-in-Publication Data
Bernstein, Jeremy, 1929–   author.
    A chorus of bells and other scientific inquiries / Jeremy Bernstein, Stevens Institute of Technology, USA.
        pages cm
    Includes bibliographical references and index.
    ISBN 978-9814578943 (alk. paper)
    1. Science--Miscellanea. 2. Quantum theory--Miscellanea. 3. Nuclear weapons--Miscellanea.
4. Financial engineering--Miscellanea. 5. Higgs bosons--Miscellanea. I. Title.
    Q173.B53 2014
    530--dc23
                                                        2013038487

British Library Cataloguing-in-Publication Data
A catalogue record for this book is available from the British Library.

Typeset by Stallion Press
Email: enquiries@stallionpress.com

Printed in Singapore

Preface

Ever since childhood I have been afflicted with something I call "grasshopper mind." I light on a subject, study it intensely, and then move on to something else. There seems to be no orderly pattern to this choice of subjects. What I do, just do, is to write an essay on these various subjects. Many of these essays I send to friends and colleagues and some I publish. This collection has a little of both. For the purposes of the collection I have gone over the essays to try to correct mistakes and to bring them up-to-date.

The first part reflects my conviction that at the present time there is no satisfactory interpretation of the quantum theory. This feeling was largely the result of my friendship with the late John Bell. In talking to Bell about the quantum theory I often felt like the respondents in the Socratic dialogues described by Plato. There were times when I understood the feelings of the Athenians who insisted that Socrates drink the hemlock. Bell was very persuasive and the title of this collection is a dedication to him.

I have for a long time been interested in the history of nuclear weapons. One of the essays describes P.A.M. Dirac's contribution to the theory of isotope separation by the use of gas centrifuges. Dirac produced an expression of the maximum amount of separation that an ideal centrifuge of given dimensions can produce. This is an analogue in a sense of Carnot's study of the most efficient possible steam engine. These limits are aspirational. At the time the memorandum of Peierls' and Frisch's was also aspirational. It described how to make an atomic bomb and this changed the world.

The two essays on quantitative financial analysis have no origin that I can recall. I have no idea how I heard the name of Louis Bachelier who at the end of the 19th century created the theory of stock analysis which used methods of Brownian motion later discovered by Einstein and much later rediscovered by the modern quantitative analysts. The other essays are my attempt to understand the contemporary financial instruments.

Finally there are three essays on the Higgs boson and the generation of mass. I wrote them both to explain the material to myself and to a non-specialist reader.

For thirty five years my night job was as a staff writer for *The New Yorker* magazine. Among other things I wrote profiles of physicists such as T.D. Lee and C.N. Yang, I.I. Rabi, Hans Bethe and Einstein. I would like to think that some of the lessons I learned in writing for that magazine also carry over to these essays.

Contents

Introduction

When I was a kid I was absorbed by comic books; things like Superman, Buck Rogers, Batman and the rest. On the back of these comic books there were frequently found advertisements for things that would lead to self-improvement. One of them was for Charles Atlas. He was a body-builder whose real name was Angelo Sicilliano. These advertisements showed a young Charles Atlas at the beach — a ninety-seven pound weakling. He was trying to chat up a girl when this bully comes along and kicks sand in his face. Atlas is too scared to do anything and he loses the girl. She probably didn't realize the possibility that he might grow up to be Bill Gates. Atlas goes home and invents a body-building method called "dynamic tension." You did things like press your hands together to build muscles. It worked — at least for Atlas — and he returned to the beach, put the bully in his place and won the girl. I then tried this for several weeks and noticed no difference.

Another advertisement offered a cure for the "grasshopper mind." This was an affliction — from which I suffered — that your focus jumped from subject to subject with no special rhyme or reason. I forget what you were supposed to do to cure yourself but again for me it didn't work. I still have it. Something triggers my interest and I become totally absorbed — for awhile at least. At some point I think that I have learned enough meaning that I have to write it down. This has produced many good essays, some of which have been published and some not. This collection contains some of them. I have published essay collections before but this one is different. There are some equations to be found. They illuminate the prose and are indispensable. I hope that I have made them easy to follow. I also hope that the essays are as enjoyable to read as they were in writing them.

The essays fall into four categories. The subject of the first deals with the quantum theory and its creators. They include profiles of P.A.M. Dirac and Max Born, for example. I discuss some work of Dirac which is not very well known. During the war he was recruited to work on the theory of isotope separation. He derived the basic equations that are still in use. This includes an expression for the maximum separation power an ideal gas filled centrifuge can have. I give a derivation of this which still plays an essential role in centrifuge technology.

My interest in the foundations of the quantum theory has a great deal to do with my friendship with the late John Bell. I knew him for some 30 years. In September of 1990 he was scheduled to give a lecture in New York which I had arranged, when his wife Mary called me from Geneva to say that John had had a stroke from which he was not expected to recover. Among many other things John taught me a simplified version of his inequality which I present. He also persuaded me to study the work of David Bohm. I explain the basics. I've also included essays on the past regarding quantum mechanics and on the measurement problem.

The articles then turn to nuclear weapons. I entered Harvard in 1947 two years after an atomic bomb was dropped on Hiroshima. What I did not know until much later was that many of my teachers helped to build the bomb. Thanks to one of them, Kenneth Bainbridge, I was offered a chance to spend the summer of 1957 at Los Alamos. At the end of the summer I was able to watch two nuclear explosions in Nevada. I have never been able to get this experience out of my mind. These essays reflect that. There is a profile of Robert Oppenheimer for example. There is also a profile of someone very few people have heard of — Gernot Zippe. He was captured by the Russians in the war and ended up in one of those scientific gulags on the Black Sea. He and his colleagues created the modern gas centrifuge. He brought it back to western Europe and it was stolen by the Pakistani metallurgist A.Q. Khan who sold it to both the Iranians and North Koreans. These are the centrifuges of which we heard much. Also, in this section I include an annotated version of the memorandum by Otto Frisch and Rudolf Peierls written in 1940, which really began our nuclear weapons program.

Then there are two essays on financial engineering. When the downturn began I felt that I should educate myself. I had no idea what a Credit Default Swap was to say nothing of LIBOR. The reader will find a primer. There is also a profile of Jean Louis Bachelier. He was the French mathematician who at the turn of the 20th century invented the whole notion of financial engineering that was redis-covered much later. It was fascinating because, without having said so, he had invented the theory of Brownian motion which was later rediscovered by Einstein.

Finally, there is a set of three essays on the Higgs boson. It seems unusual to focus so much attention on a single elementary particle but we now know why. The first of these essays was written long before any experiments were underway and deals with the underlying theory. The next essay reflects some early experimental results and asked if these were consistent with the theory. Finally there is an essay written after the discovery of the Higgs which explains it. These three essays first appeared in a slightly different form in the *American Journal of Physics* from which they were frequently downloaded so it may be worthwhile to preserve them.

The essays require different levels of mathematical sophistication to read from none to a fair amount. I hope there is something for everyone.

Part I. Quantum Theory

1. A Chorus of Bells

*"I did not dare to think that it was false, but I **knew** it was **rotten**!"*

John Bell

Not long after he matriculated at Queens College in Belfast in 1945, John Bell took his first course in quantum mechanics from Robert Sloane. At the time Bell had vivid red hair but not the beard he wore later after he scarred his lip in a motorbike accident. One pities poor Sloane. Most students, when they first encounter quantum mechanics, are in a state of shock and awe. Not Bell. He decided that at its base it was fraudulent. He had screaming arguments with Sloane. Subsequently, Bell accepted all the practical applications of quantum mechanics. He later introduced the acronym FAPP — For All Practical Purposes. He agreed that quantum mechanics was the greatest FAPP theory ever created. He was always sure that it would pass the various tests he proposed for it. But it was the muddle that he perceived in its foundations he could not stand. Take the wave function for example.

When we learn (to take an example) about the quantum mechanics of the electron in the hydrogen atom, we have, I am sure, some sort of picture of a tiny charged object whose position is described by its wave function. All of our instincts tell us that the electron

has a position which the wave function is telling us about. We must keep reminding ourselves that if we believe the interpretation of the quantum theory as expressed, say, by Bohr, then the wave function is not a description of reality. It **is** reality. As Bohr put it:

"There is no quantum world. There is only an abstract quantum physical description. It is wrong to think that the task of physics is to find out how nature is. Physics concerns only what we can say about nature."[a]

Bell found this totally unacceptable. Even more unacceptable was what he found that quantum theory — at least the usual interpretation — had to say about measurement.

In the theory there are "observables" represented by self-adjoint operators. These operators have real eigenvalues and associated eigenvectors. If the system is in a state ψ, and the observable in question is A, then we can expand ψ in a sum over the eigenvectors associated with A. The coefficients in the sum are complex numbers whose absolute squares represent the relative probabilities of measuring given eigenvalues.[b] This is an assumption which is often called "Born's rule" (after Max Born) who introduced the probability interpretation of the quantum theory. Bohr insisted there were "apparatus" and that these were necessarily described by classical — i.e., non-quantum — physics. These apparatus performed measurements on quantum systems. He was never very clear on exactly how to make this distinction except that systems were "small" and the apparatus were "large." This lack of precision drove Bell crazy and he kept referring to Bohr as an "obscurantist." FAPP there was in general no problem and it is a separation that experimental physicists make on a daily basis. We could, of course, insist that an apparatus was as quantum mechanical as anything else. But then we are

---

[a]This is quoted and discussed in *The Philosophy of Quantum Mechanics* by Max Jammer, John Wiley, New York, 1974, p. 204. It is actually something that Bohr's assistant at the time, Aage Peterson, reported Bohr as having said. For a delightful account of what Bohr did and did not say, see "What's wrong with this Quantum World" by N. David Mermin, *Physics Today*, February 2004, 10–11. Bohr said a great many things only some of which are comprehensible to me.

[b]The state vector is normalized to unity which permits this interpretation.

apparently driven into an infinite regress ending up with the experimenter's brain. On top of this there was the act of measurement itself. An actual measurement projects out one of the components of the wave function, something which cannot be described using the formalism of the quantum theory that applies to the behavior of the system up to the time when this measurement is actually recorded. What are the dynamics of this collapse? When exactly does it take place and does it require the consciousness of an "observer" to make it happen? It was over matters like this, where he had his screaming arguments with poor Doctor Sloane.

Bell was philosophically inclined even in high school. He used to bring home from the library large books of Greek philosophy. His working class parents referred to him as "the professor" — little did they know. In 1948, Born delivered the so-called Waynflete Lectures at Oxford. Soon after they were published under the title *Natural Philosophy of Cause and Chance.*[c] Bell was much taken by the lectures. However he came across the following:

> "I expect that our present theory will be profoundly modified. For it is full of difficulties which I have not mentioned at all — the self-energies of particles in interaction, and many other quantities, like [the fact that] collision cross-sections lead to divergent integrals. But I should never expect that these difficulties could be solved by a return to classical concepts. I expect just the opposite, that we shall have to sacrifice some current ideas and use still more abstract methods. A more concrete contribution to this question has been made by J. von Neumann in his brilliant book *Mathematische Grundlagen der Quantenmechanik*. He puts the theory on an axiomatic basis by deriving it from a few postulates of a very plausible and general character about the properties of 'expectation values' (averages) and their representations by mathematical symbols. The result is that the formulation of quantum mechanics is uniquely determined by these axioms; in particular no concealed parameters [hidden variables] can be introduced with the help of which the indeterministic description could be transformed into a deterministic one..."[d]

But in early 1952 the papers of David Bohm appeared. Bohm had revived an approach to the quantum theory that had first been

---

[c]A slightly more recent edition is Dover Publications, New York, 1964.

[d]Born, 1964, op. cit. p. 109.

introduced by Louis de Broglie in the late 1920s. de Broglie considered the Schrödinger wave function as describing a "pilot wave" that guided the motion of some more or less classical particles. At a meeting at which de Broglie described his scheme he was subjected to withering criticism by Pauli and he dropped the subject. It was discovered independently by Bohm some three decades later. Bohm found no difficulty in dispatching Pauli's objections. Indeed Bohm's formalism, which I will discuss shortly, can reproduce all the results of non-relativistic quantum theory in a deterministic fashion and hence is a *prima facie* counterexample to von Neumann's claim. When Bell saw this he realized that something had to have been wrong with von Neumann's claim. By this time Bell had graduated with first class honors from Queen's and had gone to work at a sub-station of the Atomic Energy Research Establishment at Malvern in Worcestershire. He was assigned to work on the design of a linear accelerator. Up to this point there had been nothing he could do about what von Neumann wrote since Bell did not read German and von Neumann's book had not yet been translated into English. But at Malvern he found a colleague named Fritz Mandl who both knew German and was interested in the foundations of the quantum theory. He translated the relevant parts of von Neumann's work.

I have read this section of von Neumann's several times and each time I am amazed that Bell could extract with such clarity the central point. Incidentally, Basil Hiley, who was a close collaborator of Bohm's, informs me that he and Bohm "puzzled over von Neumann for a considerable time but could not spot where the problem lay."[e] Von Neumann was a mathematician and a very solid one. His book with its axioms and theorems reads more like a math text than a book about physics. There is to be sure some physics involved. He presents, for example, the first accurate description of the measurement process in the quantum theory. The discussion of what von Neumann refers to as "hidden variables" appears unexpectedly towards the end of the book.[f] To understand it I prepare the reader by reminding that

---

[e]I thank Basil Hiley for this and for other comments.

[f]More exactly on page 320 of the English edition of *Mathematical Foundations of Quantum Mechanics*, Princeton University Press, Princeton, 1955.

if the state of a system is described by a wave function φ then the "expectation value" of an observable A, $\langle A \rangle$ is given by

$$\langle A \rangle = \int \varphi^* A \varphi dV.$$

In terms of this expectation value the square of the "dispersion" of this observable in this state is given by

$$(\Delta A)^2 = \langle A^2 \rangle - \langle A \rangle^2.$$

von Neumann's way of formulating the hidden variable problem is to consider what he calls "dispersion free" states, for which the above quantity is zero. If φ happened to be one of the eigenvectors of A, then as far as A was concerned this state would be dispersion free. von Neumann proposed taking an ensemble of such states and averaging over them somehow to reproduce the results of quantum mechanics. He argued that this was impossible. "There are no ensembles free of dispersion," he wrote.[g] The assumption he makes — for standard quantum mechanics it is a trivial consequence of the definition of the expectation value — is that expectation values are linear; i.e.,

$$\langle \alpha A + \beta B \rangle = \alpha \langle A \rangle + \beta \langle B \rangle$$

even if A and B do not commute, which is a remarkable result if one thinks about it. But the eigenvalues of sums of non-commuting operators are not additive. Bell's favorite example involves the Pauli spin matrices. The eigenvalues of $\sigma_x = \begin{pmatrix} 0 & 1 \\ 1 & 0 \end{pmatrix}$ are ± 1, as are the eigenvalues of $\sigma_y = \begin{pmatrix} 0 & -i \\ i & 0 \end{pmatrix}$, while the eigenvalues of the sum are $\pm\sqrt{2}$. But in a dispersion free state the expectation value of an observable must equal one of its eigenvalues, which is not true here since the eigenvalues are not additive and the expectation values are. This is certainly correct and knocks down the straw hidden variable theories that von Neumann considered, but it has absolutely nothing to do with the de Broglie–Bohm mechanics. I will henceforth refer to this as Bohmian mechanics since I will be using his formalism. I am

---

[g]von Neumann, op. cit. p. 323.

aware of the fact that he did not like this terminology but it is in common use.

In this theory, there are particles that follow classical trajectories which are determined by first-order differential equations for the particle coordinates $\mathbf{x}(t)$. I will begin by considering a single particle. As we shall see, what drives the differential equation — the "force" term — is a wave function $\psi(\mathbf{x}, t)$ where $\mathbf{x}$ is any point in space. ψ satisfies the Schrödinger equation

$$i\partial/\partial t \psi(\mathbf{x}, t) = H\psi(\mathbf{x}, t).$$

Here H is the Hamiltonian that may include a potential $V(\mathbf{x})$. To write the equation for $\mathbf{x}(t)$ we introduce the current $\mathbf{J}(\mathbf{x}, t)$

$$\mathbf{J}(\mathbf{x}, t) = 1/2\mathrm{im}(\psi^*(\mathbf{x}, t)\partial\psi(\mathbf{x}, t) - \psi(\mathbf{x}, t)\partial\psi^*(\mathbf{x}, t))$$

where m is the mass of the particle. We also introduce the density $\rho(\mathbf{x}, t)$ where

$$\rho(\mathbf{x}, t) = \psi^*(\mathbf{x}, t)\psi(\mathbf{x}, t).$$

Using the Schrödinger equation one can establish the continuity equation

$$\partial/\partial t \rho + \nabla \cdot \mathbf{J} = 0.$$

The equation for the trajectory of $\mathbf{x}(t)$ is given by — an assumption

$$d\mathbf{x}(t)/dt = \mathbf{J}(\mathbf{x}(t), t)/\rho(\mathbf{x}(t), t)$$

It is comforting to report that for a free particle, $V = 0$,

$$d\mathbf{X}(t)/dt = \mathbf{p}/m.$$

Incidentally, Bohmian mechanics is very often called a "hidden variable" theory. It seems to me that this is a misnomer. There is nothing hidden about the position variables of the particles. It would, I think, be better to call it a "classical variable" theory. The quantum mechanical features enter because given a set of initial conditions the trajectory is then determined; these initial conditions are distributed with probabilities given by $|\psi(\mathbf{x}, 0|^2$. Many examples have been worked out including the notorious double slit experiment. In Bohmian mechanics, the particle goes through one slit or the other while the guide wave goes through both, which accounts for the interference pattern.

While Bohm does discuss the "non-locality" of the theory, it was Bell who first stated this feature with clarity. I find a good deal of confusion in discussions of this so I am going to introduce the notions of "strong" and "weak" non-locality. I begin with strong non-locality. I will define this by saying that a theory that is strongly non-local has "tachyons" — particles that always move faster than light — in it. I am well aware that people who discuss this kind of non-locality often mention super-luminal signals that transport "information." This brings in a discussion of what "a signal" is and what "information" is that I want to avoid. It is well-known that tachyon theories can be made Lorentz invariant. That is not the problem. The problem is with causality. This difficulty has been known since Einstein first pointed it out in 1907.[h] If there is a faster-than-light particle that propagates between two spacetime points with the absorption event occurring later in some reference system than the emission, then it is possible to find a Lorentz transformation to a system moving less than the speed of light in which the order of these events is reversed. We would now say that in this system the absorption of the tachyon has been converted into the emission of an anti-tachyon. We can play all sorts of games with this and Bell even invented the perfect tachyon murder.[i] The perpetrator shoots the victim in one coordinate system, but to the jury in another system it looks as if an anti-tachyon has been emitted followed by the demise of the victim — no murder.

Tachyons are undesirable and Bohmian mechanics does not have them. But there is weak non-locality which is an ineluctable feature of the quantum theory. Einstein referred to it as "spooky actions at

[h] *Ann. Phys. Lpz.* **23** (1907) 371.

[i] J.S. Bell, *Speakable and Unspeakable in Quantum Mechanics*, Cambridge University Press, New York, 2004 p. 235–6.

a distance" and Schrödinger coined the term "entanglement." In the paper in which he introduced the term, he wrote:

> "When two systems, of which we know the states by their respective representatives, enter into temporary physical interaction due to known forces between them, and when after a time of mutual influence the systems separate again, then they can no longer be described in the same way as before, *viz.* by endowing each of them with a representative of its own. I would not call that *one* but rather *the* characteristic trait of quantum mechanics, the one that enforces its entire departure from classical lines of thought. By the interaction the two representatives [the quantum states] have become entangled."[j]

It is clear that any scheme that purports to reproduce the quantum theory must have this feature which I have called weak nonlocality. Bohmian mechanics does have it. This shows up when we have two particles in an interaction which has produced an entangled state. Each particle has its own differential equation driven by a common wave function. But if the particles are entangled this wave function $\psi(\mathbf{x}_1, \mathbf{x}_2, t)$ is not separable. The time $\mathbf{t}$ is common because the theory is non-relativistic. Hence the behavior of one of the particles is dependant on the instantaneous behavior of the other, however widely separated. There are no tachyons here, just entanglement. In 1966 Bell published an article in *The Reviews of Modern Physics* entitled "On the problem of hidden variables in quantum mechanics."[k] He ends it by saying, "It would be interesting perhaps to pursue some further "impossibility proofs" replacing the arbitrary axioms objected to above by some conditions of locality or of separability of distant systems." But to this there was attached a footnote which he added in the proof, that this work had at the time of the publication of the article already been done. This was of course a reference to the inequality that he had derived.

Rigorous proofs of this inequality abound and I have no intention of reproducing any of them. Instead I am going to give a poor man's

---

[j]E. Schrödinger, Discussion of Probability Relations Between Separated Systems, *Proceedings of the Cambridge Philosophical Society* **31**, 1935, p. 555.

[k]This article is reprinted in Bell 2004, op. cit. The page numbers I will cite are from this reference and in this instance, from page 11.

version which, in its outlines, was suggested to me by Bell when I asked him how he explained it to non-specialists with a limited attention span. I have "gussied up" Bell's version and I do this by introducing what I call "Einstein robots." These are incredibly smart robots that can be programmed to reproduce the results of quantum mechanics. They can be made so small that they can fit on single atoms. The one thing they cannot do is to exchange signals of any kind faster than the speed of light. No tachyon guns for them. I am going to program the robots to reproduce the Stern–Gerlach experiment. You will recall that in 1922 Otto Stern and Walther Gerlach sent beams of silver atoms through an inhomgeneous magnetic field. Much to their surprise the beam was split in two and produced two separated lines on a photographic plate. What they did not know at the time was that they had measured the spin of the electron. On the one hand, the notion of spin had not yet been invented. On the other hand, the electronic structure of silver was not yet known. The core of the silver electrons are in a net state of zero angular momentum while a single valence electron in an S state is outside. This electron spin gives the atom its net angular momentum.

The silver atoms with their attached robots are then launched in a beam with a random mixture of spin "up" and spin "down" atoms. When an atom comes under the influence of the inhomogenous magnetic field there is a force on it whose direction depends on the orientation of the spin. When the robot senses this direction it guides the atom along the appropriate orbit. This way the Stern–Gerlach pattern is reproduced. Having accomplished this with no difficulty the robot is given a new task. Now there are two magnets, one behind the other. The robot collects all the spin up events from the first magnet and guides them to the second magnet. If its field is oriented in the same direction as the first, the robot will guide all the atoms in the spin up direction. But suppose we rotate the second magnet around the direction of the incoming beam and through an angle θ. Quantum mechanics tells us[1] that with this rotation, if the spin was up in the original system, then the probabilty of finding it up in the

---

[1]See the appendix for the details.

rotated system is $\cos^2(\theta/2)$ while the probability of finding it down is $\sin^2(\theta/2)$. Hence with this rotation there will now be two lines on the photographic plate with varying intensities. When the two magnets are at right angles the intensities are the same. All of this we can teach to the robots.

Now we give the robots a new and more interesting task. We prepare two silver atoms in a spin singlet state whose wave function is symbolically $(\uparrow_1\downarrow_2 - \downarrow_1\uparrow_2)/\sqrt{2}$ where the arrows refer to the directions of the spin. This is the canonical example of an entangled state. The silver atoms fly off in opposite directions with their robots attached. They encounter two widely separated Stern–Gerlach magnets. Each robot is on its own and guides its silver atom depending on the orientation of the magnets as it has been instructed to do. If the magnets are parallel the anti-correlation of the two spins is observed. If one of the magnets is rotated through an angle $\pm\theta$ then one of the robots can be instructed for a fraction of the time proportional to $\sin^2(\theta/2)$ to guide the trajectory of the silver atom so that the two spins are measured to be in the same direction. This agrees with the quantum mechanical result (see the appendix for the details). But suppose one magnet is rotated through θ and the other through $-\theta$. Each robot will act as if it is supposed to change its orbit a fraction of the time proportional to $\sin^2(\theta/2)$ so according to the robots the total fraction of the time when the two spins are measured to be the same is proportional to $2\sin^2(\theta/2)$. But the correct quantum mechanical result is $\sin^2(\theta)$, so we are stuck. In the interval $0 < \theta < \pi/2$ we have $\sin^2(\theta) > 2\sin^2(\theta/2)$. This is a primitive example of a Bell inequality.[m] Quite generally, no local hidden variable theory can reproduce all the results of quantum mechanics.

---

[m]The purpose of this footnote is to remind the reader of, or introduce the reader to, Bell's original inequality which he published in *Physics* **1** (1964) 195–200. The context is again a double Stern–Gerlach experiment. Let **a** be the direction of one magnet and **b** the direction of the other. Let λ be some "hidden variable." There might be several but one will do. The result of a measurement with the A magnet is given by $A(\mathbf{a}, \lambda) = \pm 1$ while the result of a measurement with the B magnet is $B(\mathbf{b}, \lambda) = \pm 1$. The locality is represented by the fact that A is only a function of **a** and B is only a function of **b**. The correlation of these measurements is given by a function $P(\mathbf{a}, \mathbf{b})$ which a weighted integral over λ with a weight function

Having spoken to Bell about all of this, I am quite sure that he believed that any experiments done on his inequalities would agree with quantum mechanics. Quantum mechanics gives correct results in domains as widely separated as superconductivity and supernovae. It would be somewhat absurd to think that it would break down in a Stern–Gerlach experiment — and indeed it didn't when it was tested by people such as Alain Aspect. Bell once said to me with some regret that it showed that Einstein was wrong and Bohr was right. Einstein, he felt, was acting like a reasonable scientist while Bohr was an obscurantist. "The reasonable thing," he said, "just doesn't work." I do not fully understand what Einstein wanted. As a guess I think he wanted to see quantum mechanics emerge from some underlying deterministic theory in somewhat the same sense that thermodynamics emerges from statistical mechanics. He no doubt wanted the underlying theory to be local, or free of spooky actions at a distance. What Bell showed is that the underlying theory, if there is one, cannot be local. We know Einstein's feelings about Bohmian mechanics. He expressed them in a letter to Born dated May 12, 1952:

> "Have you noticed that Bohm believes (as de Broglie did, by the way 25 years ago) that he is able to interpret the quantum theory in deterministic terms? That way seems too cheap to me. But you, of course, can judge this better than I."[n]

I wish I knew what Einstein meant by "cheap" in this context.

$\rho(\lambda)$; i.e.,

$$P(\mathbf{a}, \mathbf{b}) = \int \rho(\lambda) A(\mathbf{a}, \lambda) B(\mathbf{b}, \lambda) d\lambda.$$

The quantum mechanical result which is derived in the appendix is given by

$$P(\mathbf{a}, \mathbf{b})_{qm} = -\cos(\mathbf{a} \cdot \mathbf{b}).$$

Bell asked is it possible to reproduce this answer with any choice of the functions that enter the integral above. Bell derived the inequality below where **c** is a third direction

$$1 + P(\mathbf{b}, \mathbf{c}) \geq |P(\mathbf{a}, \mathbf{b}) - P(\mathbf{a}, \mathbf{c})|.$$

He showed that $P(\mathbf{a}, \mathbf{b})_{qm}$ cannot satisfy this inequality for all choices of direction.

[n]The Born–Einstein Letters, Walker and Company, New York, 1971, p. 192.

When I began to learn quantum mechanics around 1950 there were not that many texts available. One of the standard ones was *Quantum Mechanics* by Leonard Schiff. It was essentially a more detailed write up of the lectures Robert Oppenheimer had given for many years at Berkeley and Caltech. It is a good text from which to learn how to solve problems, but it contains nothing on what we would now call the foundations of the theory. The same thing is true of Dirac's masterful *The Principles of Quantum Mechanics*. In the first chapter Dirac states that a measurement collapses the wave function and that is that. He once remarked to someone that he thought that it was a good book but that the first chapter was missing. And then in 1951 Bohm published his text *Quantum Theory*.[o] It is full of discussion of the foundations. Abner Shimony, who made very basic contributions to the development of Bell's inequalities, asked his then thesis advisor Eugene Wigner what he thought of the book. Wigner told him that it was a good book except that there was too much "schmoozing." The schmoozing is just what I liked since it dealt with the foundations of the theory. What is remarkable about the book is that it contains a "proof" that the results of the quantum theory cannot emerge from hidden variables. He wrote, "We conclude that no theory of mechanically determined hidden variables can lead to *all* of the results of the quantum theory." But not long after the book was published he had produced a theory which did just that. One of the things that Bell took from the book was Bohm's novel presentation of the Einstein, Podolsky, Rosen experiment which they first published in 1935,[p] This version of the EPR experiment has been with us every since. The ingredients will be familiar.

Some mechanism produces a pair of spin-1/2 particles in a singlet state. They fly off in opposite directions to a pair of Stern–Gerlach magnets. Let us say that one of the magnets is oriented in the z-direction and let us say that it measures the spin of one of the particles to be "up." Because of the correlation we have already

---

[o] Prentice Hall, Englewood, New Jersey.

[p] "Can Quantum Mechanical description of physical reality be considered complete?" A. Einstein, B. Podolsky and N. Rosen, *Physical Review* **47**, 696 (1935).

discussed we would predict that, when measured, the spin of the other particle will be "down." EPR goes a step further. They argue that in this set up the z-component of the spin of the other particle has been implicitly measured and that this implicit measurement has conferred "reality" on this quantity. One can then set about to measure the x-component by rotating the magnet. This having been done we have both components measured which quantum mechanics says is impossible. The solution to this problem, if it is a problem, is to insist that "implicit measurements" in the quantum theory don't count. Either you measure something or you don't. You cannot measure the x and z components simultaneously. You need two different experiments. Bell of course understood this, but I think that it was thinking about double Stern–Gerlach experiments in this context that set him off.

In the spring of 1984 I decided that I would try to write a *New Yorker* profile of Bell whom I had known since he first went to CERN in 1960. We had a new editor at the *New Yorker*, Robert Gottlieb, who did not seem to have that much interest in science, but since I was going to CERN anyway on some leave there was not much to lose. Bell seemed agreeable and over some days I interviewed him on tape. Later I wrote my profile which was turned down. I published it in a 1991 collection titled *Quantum Profiles*.[q] By the time the book came out John had died. He died on October 1, 1990 of a cerebral aneurism. He had been nominated for a Nobel Prize which I think he would have won. He had also become something of a cult figure especially among New Age types who had no real understanding of what he had done. John seemed to accept all of this with a wry amusement. In 1979 he even attended a meeting organized by the Maharishi Mahesh Yogi, who had in fact been a physics major, which took place at the Maharishi university above Lake Lucerne. Bell told me that while he found the occasion rather absurd he nevertheless liked the vegetarian meals. During my interviews I got the impression that none of the formulations of the quantum theory really satisfied him. I think the de Broglie–Bohm came closest although he was bothered

---

[q]Princeton University Press, Princeton.

by making it Lorentz invariant. He said that someday he might write a book about all of this. He never did.

Appendix: Spinning[r]

In the body of the text I mentioned some of the consequences of rotating the Stern–Gerlach magnets. In this appendix I want to fill in the details. We imagine first performing measurements of the spin along the z-axis when the particles are moving in the y direction. We then rotate the magnet through an angle θ in xz plane. The Pauli matrix which was $\left(\begin{smallmatrix} 1 & 0 \\ 0 & -1 \end{smallmatrix}\right)$ is in the new system $\left(\begin{smallmatrix} \cos(\theta) & \sin(\theta) \\ \sin(\theta) & -\cos(\theta) \end{smallmatrix}\right)$. This matrix has the eigenvectors $\left(\begin{smallmatrix} \cos(\theta/2) \\ \sin(\theta/2) \end{smallmatrix}\right)$ and $\left(\begin{smallmatrix} -\sin(\theta/2) \\ \cos(\theta/2) \end{smallmatrix}\right)$. We can expand the vector $\left(\begin{smallmatrix} 1 \\ 0 \end{smallmatrix}\right)$ in this basis and write $\left(\begin{smallmatrix} 1 \\ 0 \end{smallmatrix}\right) = a_+ \left(\begin{smallmatrix} \cos(\theta/2) \\ \sin(\theta/2) \end{smallmatrix}\right) + a_- \left(\begin{smallmatrix} -\sin(\theta/2) \\ \cos(\theta/2) \end{smallmatrix}\right)$. Which implies that $a_+ = \cos(\theta/2)$ and $a_- = -\sin(\theta/2)$. This means that the probability of finding the spin up in rotated magnet is $\cos^2(\theta/2)$ while the probability of finding spin down is $\sin^2(\theta/2)$. Hence with the entangled singlet particles, if I say measure spin down (or up) in one magnet then the probability of measuring the same result in the rotated magnet is $\sin^2(\theta/2)$ while the probability of measuring the opposite spin is $\cos^2(\theta/2)$. Thus the quantum mechanical correlation is given by

$$\sin^2(\theta/2) - \cos^2(\theta/2) = -\cos(\theta).$$

Can we program the robots to reproduce this? There is no problem programming a robot when it finds the rotated magnet to alter its trajectory so that the two spins are aligned $\sin^2(\theta/2)$ fraction of the time agreeing with quantum mechanics. But if both magnets are rotated in opposite directions by the same angle then the robots will alter their trajectories so that agreement occurs $2\sin^2(\theta/2)$ of the time. But the quantum prediction is that agreement in this case occurs $\sin^2(\theta)$ percent of the time in the range $0 < \theta < \pi/2$.

---

[r] I am very grateful to David Mermin for the critical remarks on an earlier draft that inspired me to write this appendix. I am also grateful to Elihu Abrahams for a critical reading of this draft.

$\sin^2(\theta) > 2\sin^2(\theta/2)$ as shown in the figure below. This is Bell's inequality in this simple case.

The upper blue line is the plot for $\sin^2(\theta)$ and the lower red line is the plot for $2\sin^2(\theta/2)$.

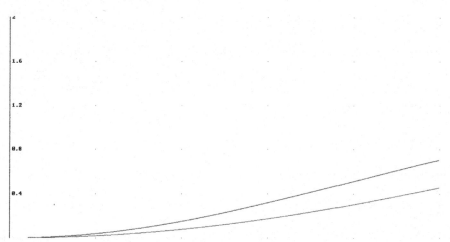

2. A Quantum Past

"I deduce two general conclusions from these thought-experiments. First, statements about the past cannot in general be made in quantum-mechanical language. We can describe a uranium nucleus by a wave-function including an outgoing alpha-particle wave which determines the probability that the nucleus will decay tomorrow. But we cannot describe by means of a wave-function the statement, 'This nucleus decayed yesterday at 9 a.m. Greenwich time.'

As a general rule, knowledge about the past can only be expressed in classical terms. My second general conclusion is that the 'role of the observer' in quantum mechanics is solely to make the distinction between past and future. The role of the observer is not to cause an abrupt 'reduction of the wave-packet,' with the state of the system jumping discontinuously at the instant when it is observed. This picture of the observer interrupting the course of natural events is unnecessary and misleading. What really happens is that the quantum-mechanical description of an event ceases to be meaningful as the observer changes the point of reference from before the event to after it. We do not need a human observer to make quantum mechanics work. All we need is a point of reference, to separate past from future, to separate what has happened from what may happen, to separate facts from probabilities."

Freeman Dyson[a]

---

[a]Thought Experiments in Honor of John Archibald Wheeler, in *Science and Ultimate Reality*, Cambridge University Press, New York, 2004, p. 89.

A long time ago Murray Gell-Mann told me that Feynman had once said to him that quantum mechanics could not account for history. I think that this is another way of stating Dyson's point. The first thing I want to do in this essay is to explain my understanding of what this means. I am in these matters very conscious of my favorite aphorism of Niels Bohr. He cautioned us not to speak — or write — more clearly than we think. In any event the first question to be answered is what is understood by "quantum mechanics." In this I shall frequently employ John Bell's acronym FAPP — For All Practical Purposes. The quantum mechanics I shall refer to here is FAPP mechanics. It is the quantum mechanics found in Dirac's book. If you haven't read his book for a while, I suggest you look at the opening chapter. There is not one word about any of the foundational issues that have become fashionable. You will find no reference to the Einstein, Podolsky, Rosen paper. The word "complementarity" is not found, nor is "entanglement", nor "non-locality." There is no measurement "problem" but only a matter of fact statement of what a measurement is. John Bell once told me that Dirac had said to a colleague that it was a good book, but that the first chapter was missing. As far as the book is concerned, his only interest is how to formulate the theory in order to solve problems. In short, it is pure FAPP.

In FAPP language we have a quantum mechanical system described by a wave function $\psi(t)$. I am only interested in the time variable. The wave function obeys a Schrödinger equation with a Hamiltonian H. The formal solution to this equation is $\psi(t) = \exp(iHt)\psi(0)$. Throughout I am setting $\hbar = 1$. Thus to recover $\Psi(0)$ from $\psi(t)$ all we have to do is to multiply by $\exp(-iHt)$. Haven't we then recovered the past? What is all the fuss about? The problem is that there is more to life than the wave function. There are the "observables" which represent what we really want to know about the system. These observables are described by Hermitian operators A, B, C and so on. We can expand ψ in a sum over the orthonormal eigenfunctions of any of these operators. The coefficients in the expansion are related to the probabilities that in a measurement the system will be found to have one of these eigenvalues. This is "Born's rule" and in FAPP it must be assumed. To find which of these eigenvalues the system actually has, we must perform a measurement. Stripped to its essence the apparatus that produces this measurement projects out from the sum of eigenfunctions one of them. After the measurement the rest of the terms in the sum disappear. Using the term of art, the wave function "collapses." It is at this point that we lose our capacity to reconstruct the past. Projection operators are singular. They do not have inverses. All the king's horses and all the king's men cannot put the wave function back together again. It was von Neumann in the early 1930's who first noted that in FAPP mechanics there were two kinds of processes. There were processes that could be described by a Schrödinger equation and there were measurements which could not. He did not, as far as I know, comment on what this implied for retrodiction. A case in point is an electron described by a spherically symmetric Schrödinger wave. If this electron strikes a detector it does so at a place — a spot. After this happens all trace of the spherically symmetric wave function vanishes.

I have certainly not made a careful search of the literature but among the founding fathers of FAPP I can come up with only two references that deal with the matter of the quantum past. One is Heisenberg and the other is a paper by Einstein, Richard Tolman, and Boris Podolsky, "Knowledge of Past and Future in Quantum Mechanics" which they wrote in 1931 when Einstein was spending time at Caltech. I think that this was Einstein's first paper in English.[b] But first, Heisenberg.

In 1929 Heisenberg gave a series of lectures on the quantum theory at the University of Chicago. These were published in 1930 in a book entitled *The Physical Principles of the Quantum Theory*.[c] There is one paragraph devoted to the quantum past which I will quote in its entirety.[d]

"The uncertainty principle refers to the degree of indeterminateness in the possible present knowledge of the simultaneous values of various

---

[b] *Phys. Rev.* 37 (1931) 78.

[c] Dover Press, New York, 1930.

[d] Heisenberg, op. cit. p. 20.

quantities with which the quantum theory deals; it does not restrict, for example, the exactness of a position measurement alone or a velocity measurement alone. Thus suppose that the velocity of a free electron is precisely known, while the position is completely unknown. Then the principle states that every subsequent observation of the position will alter the momentum by an unknown and undeterminable amount such that after carrying out the experiment our knowledge of the electronic motion is restricted by the uncertainty relation. This may be expressed in concise and general terms by saying that every experiment destroys some of the knowledge of the system which was obtained by previous experiments."

Then he writes,

"This formulation makes it clear that the uncertainty relation does not refer to the past: if the velocity of the electron is at first known and the position then exactly measured the position for times previous to the measurement may be calculated. Thus for the past times $\Delta \times \Delta p$ is smaller than the usual limiting value, but this knowledge of the past is of a purely speculative character, since it can never (because of the unknown change in momentum caused by the position measurement) be used as an initial condition in any calculation of the future progress of the electron and thus cannot be subjected to experimental verification. It is a matter of personal belief whether such a calculation concerning the past history of the electron can be ascribed any physical reality or not."

Clearly the electron had a past history but what Heisenberg seems to be saying is that it cannot be described by quantum mechanics.

The Einstein, Tolman, Podolsky paper makes a somewhat different point. They produce a thought experiment which I will describe in which two particles, electrons say, move along two trajectories one long and one short to a common detector. They argue that if retrodiction were allowed for the particle that arrives first then you could make a prediction about the arrival of the second particle that would violate the uncertainty principle. The conclusion is that the uncertainty principle applies to reconstructions of the past as well as predictions of the future. On its surface this seems to be inconsistent with everything we believe about the past. I may conjecture that the Sun will come up tomorrow, but I know that the Sun came up yesterday. Here is how they begin their brief paper.

"It is well-known that the principles of quantum mechanics limit the possibilities of exact predictions as to the future path of a particle. It has sometimes been supposed, nevertheless, that the quantum mechanics would permit an exact description of the past path of a particle."

One would like to know who "supposed" this. There are no references of any kind in their note. They go on,

"The purpose of the present note is to discuss a simple ideal experiment which shows that the possibility of describing the past path of one particle would lead to predictions as to the future behavior [sic] of a second particle of a kind not allowed in the quantum mechanics. It will hence be concluded that the principles of quantum mechanics actually involve an uncertainty in the description of past events which is analogous to the uncertainty in the prediction of future events. And it will be shown for the case in hand, that this uncertainty in the description of the past arises from a limitation of the knowledge that can be obtained by measurement of momentum."[e]

In this setup the authors imagine a box on some sort of scale. Inside the box are particles in agitation. There are two holes and a shutter that opens and closes them. Einstein buffs will be reminded of a similar apparatus that Einstein introduced in the 1930 Solvay meeting. It was the one he used to "refute" the Heisenberg energy time uncertainly relation. It will be recalled that Bohr pointed out that Einstein had left out the gravitational time variation of the clock in the box as it changed positions in the gravitational field of the Earth. Once this was taken into account the uncertainty principle was saved. In their paper the authors insist that the clock of the observer at the end of the shorter path is far away enough so that it will not be perturbed by any gravitational effects due to the weighing of the box. Once bitten twice shy. The shutter is opened briefly and out fly two particles one of which takes the short route and one of which takes the long route to the detector. The box is weighed before and after the particles are released and thus the total energy of the two particles is known. The observer at the detector measures the momentum of the particle arriving by the short route and its time of arrival. We now know the energy, and the speed of the first particle. We also know how far it has gone since we have measured that beforehand. Thus we can apparently say at what time the shutter opened and

---

[e]Einstein *et al.*, op.cit. p. 780.

how much energy the second particle has. Hence we might argue that since we know its speed and the distance it has to travel, we can say exactly at what time it will arrive at the detector and with what energy thus violating the uncertainty principle. The flaw in all of this, as the authors point out, is the assumption, reasonable on classical grounds, that we can determine the momentum of the first particle along the trajectory by a retrospective argument. If we want to be consistent quantum mechanically we must say that the particle has no momentum until we measure it, and that once we measure it its future momentum is uncertain. The authors conclude:

> "It is hence to be concluded that the principles of quantum mechanics must involve an uncertainty in the description of past events which is analogous to the uncertainty in the prediction of future events. It is also to be noted that although it is possible to measure the momentum of a particle and follow this with a measurement of position, this will not give sufficient information for a complete reconstruction of its past path, since it has been shown that there can be no method for measuring the momentum of a particle without changing its value..."[f]

What I find remarkable is that two years later Einstein was in Princeton working with Podolsky and Rosen on the inability of the quantum theory, as they saw it, to include all elements of reality. The elements of reality that they say are not included seem to me small beer as compared to the entire past! *Eppur si muove.* The Earth does move. There is a past. Hitler is dead. There was a total eclipse of the Sun on May 14, 1230. Yet FAPP mechanics cannot describe this without uncertainties. On this, as far as I can see, Einstein was silent.

There are various attitudes we can adopt towards this. The FAPP attitude was well-summarized by Alfred E. Newman. "What, me worry?" After all, as scientists we are concerned with making predictions. Leave the retrodictions to the historians. Of course there is a good deal of scientific enterprise devoted to using quantum mechanics to estimate things like the amount of helium produced in the first three minutes after the Big Bang. That appears to be retrodiction big

---

[f]Einstein *et al.* op.cit. p. 781.

time. But let us analyze the situation. It is very likely true that the experimental apparatus needed to measure this cosmological helium requires quantum mechanics to understand it. But this has nothing to do with the issue. When we predict the amount of helium that should be observed we put ourselves in a pre-helium production time frame. From this point of view the production of helium is in the future and it is to this future that we apply quantum mechanics.

We could simply accept the fact that according to FAPP mechanics the uncertainty principle applies to the past as well as the future. After all, in principle, the uncertainty principle affects everything we do. I am sure that it affects the trajectory of my bicycle commute, but FAPP, that is the least of my worries. However, accepting this does not deal with the quantum measurement issue. After such a measurement FAPP theory tells us that part — indeed most — of the wave function disappears along with our knowledge of the past. That is something to think about. It is clear that if we take this seriously we have to go beyond FAPP.

It seems to me that any interpretation of the quantum theory that addresses this must have the feature that measurements are simply just another interaction like the rest. Von Neumann's notion that there were two classes of interactions — one whose time evolution could be described by a Schrödinger equation and one of which couldn't — has to be abandoned. I will discuss two proposals for doing this each of which has adherents as well as detractors. On the one hand I am going to discuss what I will call "Bohmian mechanics" — a term which David Bohm, who invented this approach, apparently did not like. As far as he was concerned, he was just doing quantum mechanics but in a different way. However nearly everyone else calls it Bohmian mechanics — so will I. On the other hand, I am going to discuss the "decoherent history" interpretation which Murray Gell-Mann and Jim Hartle have done the most on. Sometimes this is called the "many worlds" interpretation, but not by them. I think that the term "many worlds" is misleading. As far as we know there is one world, the one we live in. First, Bohmian mechanics.

To simplify things I will restrict myself to a single particle interacted on by an external force with a potential $V(x)$. In Bohmian mechanics this particle has a real classical position $X(t)$. There are no uncertainties here. We have discussed how this works. The particle follows a classical trajectory which is determined by the Schrödinger equation. It is not my purpose here to give many examples of Bohmian mechanics in action. But let me describe one that I think is very impressive. That would be the Bohmian analysis of the two-slit experiment. I think that we all remember that when we were first told about it the question that immediately came to mind was how does the electron "know" that the other slit is open or closed? Does the electron somehow go through both slits at once? We are FAPPed into submission by being told that this is a question that FAPP mechanics cannot and will not answer. If we want to observe the electron at one slit we will destroy the interference pattern. Bohmian mechanics does away with all this FAPPness. The electron goes through only one slit while the guide wave goes through both. This is why single electrons can produce the diffraction pattern when they are let through the slits one at a time. There is no collapse of the wave function or anything like it. Likewise in a measurement there is no collapse of the wave function. For example, suppose you have an observable with two possible values. Then depending on the setting of the detector the particle will be guided to one trajectory or the other. After registering its arrival the particle can continue on. The wave function has not collapsed but the part appropriate to the other possible trajectory "decoheres" — separates out — and can no longer influence the path of the particle. All of this can be run backwards in time to re-create history. One may wonder where in all of this are the uncertainties of the quantum theory. To get a feel for what is going on, let's consider the S-wave particle I discussed earlier. The S-wave is symmetrically spread over space. But when the particle it pertains to encounters a detector the wave function, according to FAPP, collapses and the particle is located at some point in space. In Bohmian mechanics this particle has and always has had a trajectory. What then accounts for equal likelihood that particles emitted one after the other can land with equal probability in any direction? If we trace the trajectories back to their initial conditions we see that which trajectory we are on depends on the initial values of the wave function. If these are distributed statistically then so will the trajectories. While it is true in

Bohmian mechanics that each trajectory is perfectly deterministic, which one we are on is statistically determined.

People have raised various complaints about Bohmian mechanics. A frequent one is that the game of mathematical complexity is not worth the candle of the determinstic interpretation. I do not mind the mathematical complexity so long as there is someone else willing to do the mathematics. A more interesting complaint has to do with the "non-locality" of the theory. This only begins to manifest itself when there is more than one particle involved. Let us consider two particles which are interacted on with a common potential $V(x_1, x_2)$. The solution to the Schrödinger equation is a function $\psi(x_1, x_2, t)$. There is a common time because we are dealing non-relativistically. When we use this wave function in the equations of motion of the two particles the density ρ is common but the current J is different for each particle since the gradient that defines it is taken with respect to a different variable. Once the particles are in interaction the wave function cannot be written as a product of the wave functions of the individual particles. The particles have become "entangled." This means that determining the trajectory of one particle requires instantaneous input from the other no matter how far apart the particles are. It is in this sense that the theory is non-local. If Bohmian mechanics is to agree with ordinary quantum mechanics this kind of non-locality is to be expected. It is well-known that entangled particles produce instantaneous effects on correlated measurements. This is what Einstein referred to as "spooky action at a distance." No one claims that this is a violation of the theory of relativity here, and there is also no violation in Bohmian mechanics. No information bearing signals are exchanged superluminally.

Bohmian mechanics and the decoherent history interpretation have certain commonalities, the most significant of which is that neither accepts the notion that measurements differ in any way from any other kind of interaction. There is no collapse of the wave function. As formulated by Hartle a "history" is a series of answers to "yes"–"no" questions. Is the Moon at such and such a place in its orbit or isn't it? This is a question we can ask over and over again in the course of time and construct a history. To represent this mathematically we correlate to each "yes" answer a projection operator and to each "no" answer an orthogonal projection operator. These projection operators evolve in time according to the Heisenberg equation

$$P(t) = \exp(iHt)P(0)\exp(-iHt).$$

To construct a "history" we let a product of these projection operators act at a sequence of times on an initial state $|\psi\rangle$. Following Hartle[g] let us call this sequence of projection operators Cα. We make this subscript distinction because there are of course many other possible histories. How probable is this one? By assumption — this is not proved in the many histories interpretation any more than the Born rule is proved in ordinary quantum mechanics — the probabiity $P(\alpha)$ is given by

$$P(\alpha) = \|C\alpha|\psi\rangle\|^2.$$

Here we must be careful. Not every history has a well defined probability. There can be quantum mechanical interference between histories. In the two slit experiment, for example, to get the correct probability for the electron to reach a place on the detector you must take the square of the sum of the probability amplitudes and not the sum of the squares. In Bohmian mechanics this issue does not arise since each electron that passes through a slit has a classical trajectory. The histories to which one can attach a probability are "decoherent." This means that $\langle\psi|C_\beta\dagger C_\alpha|\psi\rangle$ is approximately zero onless $\alpha = \beta$. Since we do not require that it be exactly zero this kind of history is "quasi-classical." In Bohmian mechanics the trajectories are really classical. It is argued that for macroscopic objects — the Moon for example — decoherence occurs because of all the collisions between the object and, say, the microwave radiation left over from the Big Bang. In the many histories interpretation what indeed is history?

At first sight this might seem to be obvious. All we have to do is to run the chain backwards. Yes this gives one history but there are others, possibly very many others. The reason is that if all we know

---

[g]See for example: James B. Hartle, *Quantum Pasts and the Utility of History*, arXiv:gr-gc/9712001 vl 2 Dec 1997.

is the present state vector there are many paths by which we could have arrived there depending on which initial state vector we started from. We have no way of knowing this from the data we have at hand. Let us take an example discussed by Hartle — the Schrödinger's cat. I can't resist noting that when I spent an afternoon with Schrödinger in his apartment in Vienna there was no cat. In any event this unfortunate feline is put in a box that contains a capsule of poison gas and a sample of uranium. The capsule is triggered so that if the uranium has an alpha decay, the alpha sets off the trigger and the unfortunate feline expires. After a time interval we open the box and happily the cat is alive. It could, according to the many history approach, have arrived at this state in two ways. The initial state might have been a cat alive state or it might have been a coherent sum of a cat alive and a cat dead state. From the presence of the living cat we cannot decide. The vision of the past given by the decoherent history interpretation and the Bohmian seems radically different. In Bohmian mechanics we could in principle follow all the cat molecules backwards in time and arrive at one and only one past.

I don't know how you feel, but the ambiguity of the past makes me queasy. It might be entertaining to imagine that in an alternate past my grandmother who was born in a Polish shtetl could have been Eleanor Roosevelt. I readily accept that these pasts do not communicate but there seem to be too many of them from the point of view of economy. A trip to a barber wielding Occam's razor seems warranted. In any case when it comes to quantum pasts, as Duke Ellington taught us, "Things ain't what they used to be."

3. A Double Slit

"We chose to examine a phenomenon which is impossible, *absolutely* impossible, to explain in any classical way, and which has in it the heart of quantum mechanics. In reality it contains the only mystery."

Richard Feynman[a]

". . . The particle world is the dream world of the intelligence officer. An electron can be here or there at the same moment. You can choose. It can go from here to there without going in between; it can pass through two doors at the same time, or from one door to another by a path which is there for all to see, until someone looks, and then the act of looking has made it take a different path. Its movements cannot be anticipated because it has no reasons. It defeats surveillance because when you know what it is doing you can't be certain where it is, and when you know where it is you can't be certain what it's doing: Heisenberg's uncertainty principle; and this is not because you're not looking carefully enough, it is because there is no such *thing* as an electron with a definite position and definite momentum; you fix one, you lose the other, and it's all done without tricks, it's the real world, it is awake."

Tom Stoppard, *Hapgood*[b]

In the fall of 1947, I entered Harvard College as a seventeen year old freshman. I had no real ideas about my future except that I was quite sure that it would have nothing to do with science. I had no

---

[a]R.P. Feynman, R.B. Leighton and M. Sands, *1963 Feynman Lectures*, Vol. 3, Ch. 37.

[b]*Tom Stoppard Plays*, Faber and Faber, London, 1999, p. 544.

interest in science. But at that time the president of the university was James Bryant Conant, a chemist who had played a very important role in the creation of nuclear weapons. This experience had left him with a good deal of concern about this new force in the world. One manifestation of which was the notion that every Harvard graduate should have some sort of scientific education. He knew, of course, that most of us would not have careers in science, but we might well have careers where we would have to make decisions that involved science. So under the rubric of "General Education" a set of natural science courses had been instituted which presented non-mathematical surveys of different scientific fields. Every Harvard graduate was required to pass one of them unless he — there were no she — was a science major. You also had to be able to swim two laps of a twenty-five yard long pool. Faced with the former obstacle, I consulted the Harvard Confidential Guide to Courses — a small booklet produced and sold by the undergraduate newspaper, the *Harvard Crimson* — to find out which course was considered the easiest. The vote among the undergraduates who took the various courses and compared notes was unanimous that it was Natural Sciences 3, taught by the late historian of science I. Bernard Cohen.

In the interests of full disclosure I have to admit that I had had a physics course in high school. I have searched my memory to recall anything that I had learned in this course and have come up blank. It was my only contact with physics in high school. I also took the required mathematics courses in which I was a good student. If someone had told me that there were professional physicists and mathematicians who did this sort of thing for a living I would not have believed it. I had never met such a person. I had of course heard of Einstein but I had no idea of what he actually did. I also had no scientific curiosity. I tell you these things to explain what kind of student I was when I enrolled in Natural Sciences 3.

Cohen was a fine lecturer for this level of student. There were well over a hundred of us. He had a rich deep voice and a round reassuring handwriting when he wrote on the blackboards. We began with the Greeks, worked our way to Copernicus, Galileo and then Newton. Cohen was a Newton scholar so we learned about Newton's life. For the first time I realized what a very strange man he had been. His masterwork, the *Principia* which he wrote at the end of the 17th century was presented in a way, deliberately, so that it would be accessible only to scholars. It was written in Latin and used geometric arguments which could have been replaced by much simpler demonstrations using the calculus — which Newton had invented. Before it was published Newton got into another nasty priority fight with Robert Hooke as to which one of them first enunciated the correct law of gravitation. They had previously clashed on their work on light. On the other hand, Newton's *Opticks* which was published in 1704, is a very different kind of book.

In the first place it is written in English and the style is as congenial as Newton was capable of. It deals a great deal with the optical experiments Newton performed including the one with the prism that showed that white light was really a mixture of light of many colors. Unlike the *Principia* where Newton avoided speculation about the ultimate nature of things, in the *Opticks* he does speculate. There is the famous Query 31 where he talks about the atomic theory. He writes,

> "All these things being considered, it seems probable to me, that God, in the Beginning, form'd, Matter in solid, massy, hard, impenetrable Particles, of such Sizes and Figures, and with such other Properties and in such Proportion to Space, as most conduced to the End for which he form'd them, and that these primitive Particles being Solids, are incomparably harder than any porous Bodies compounded of them; even so hard, as never to wear or break in pieces; no ordinary Power being able to divide what God himself made one in the first Creation. While the Particles continue entire, they may compose Bodies of one and the same Nature and Texture in all Ages. But should they wear away, or break in pieces, the Nature of Things depending on them, would be changed. Water and Earth, composed of old worn Particles and Fragments of Particles, would not be of the same Nature and Texture now, with Water and Earth composed of entire Particles in the Beginning. And therefore, that Nature may be lasting, the Changes of corporeal Things are to be placed only in the various separations and new Associations and Motions of these permanent Particles, compound Bodies being apt to break, not in the midst of solid Particles, but where these Particles are laid together, and only touch in a few points..."

It is clear that when it came to material objects, Newton was an atomist *pure et dure*. But what about light? Here it seems to me that he was somewhat more nuanced. He writes, "In the first Books of these Opticks, I proceeded by the Analysis to discover and prove the original Differences of the Rays of Light in respect of Refrangibility, Reflexibility, and Colour, and their alternate Fits of easy Reflexion and easy Transmission, and the properties of Bodies, both opaque and pellucid on which their Reflexions and Colours depend. . . ." I am not sure what meaning is to be attached to "alternate Fits of easy Reflexion and easy Transmission" but it does not sound like a simple particle picture to me although this is how Newton was later interpreted. Opposed to this view was, in the first instance, that of Hooke who claimed that light was a wave. But of greater significance was the work of Christian Huygens — or Hargenz — the great Dutch contemporary of Newton whom he came to visit in 1689, before Newton's *Opticks* had been published. I do not know if they discussed the nature of light. Huygens own book was published in 1690. He proposed that light was an oscillation of waves in an aetherial medium. What he did not do was to consider the interference of waves. When two waves meet they can enforce each other or interfere with each other destructively. Huygens "wave" were actually pulses.[c] Nonetheless, either concept of light could account for the observations that were then available. Everything changed with the work of the late 18th and early 19th century British polymath Thomas Young. It is to him that we owe the idea that light waves can interfere with each other.

Over the years I have encountered a certain number of people that I would label "geniuses." I had an afternoon tea with Schrödinger in his apartment in Vienna and had a chance to show Dirac around that city. When I was in high school I spent some time with Duke Ellington. I had one memorable lunch with W.H. Auden and had the chance, when I was a Harvard undergraduate, to ask John von Neumann a question. The question was whether he thought the computing machine would ever replace the human mathematician. He response was, "Sonny, don't worry about it." What these people

---

[c] I am grateful to the Newton scholar Alan Shapiro of the University of Minnesota for clarifying remarks on Newton, Huygens and Young.

had in common was their ability to do things with an apparent ease that most of us think we cannot do. By this standard, or any other, Thomas Young was a genius.

Young was born into a Quaker family in Milverton in Somerset in 1773. By the age of fourteen he knew not only Latin and Greek, but had a grounding in French, Italian, Hebrew, Chaldean, Syriac, Samaritan, Persian, Turkish, and Amharic. Later in life he made basic contributions to the decipherment of Egyptian hieroglyphics, something that was completed by the French linguist Jean Françoise Champollion in 1822. Young, who became independently wealthy, was trained and practiced as a London physician and made some important contributions to the practice of medicine. This is not what made him immortal to physicists. Prior to the turn of the century he began making experiments in physics. The ones that he made on light were summarized in his November 1801 Bakerian lecture — "On the theory of light and colours" — before the Royal Society of which he had been a fellow since 1794. It was these experiments that persuaded most physicists that Huygens was right — light is a wave.

The experiment that he described in his Bakerian lecture is not the one that is usually explained to students. That came later. In this one, which was actually first done by Newton, he tells us that he made a square hole in a card. The breadth of the hole which he notes was "66/1000" of an inch. To this hole he attached a hair that ran from top to bottom. It had a diameter of 1/600th of an inch. Next he shined candle light on this arrangement. Nowadays we would use lasers which produce a concentrated beam. The hair split the light beam into two parts. I think of those bicycle races where the stream of riders is split in two by a traffic island in the middle of the road. We know that after they pass the island the two streams re-unite and to all appearances look as they did before. This is what the particle theory of light would predict. But this is not what Young observed. On the other side of the hair, light and dark fringes were observed. Using his candlelight these were red fringes. Young understood immediately the implications of this. It had to do with the interference of light waves. Later he repeated this experiment with two pinholes instead of the hair and got similar results. This was the Young's "double slit" experiment.

While all of this was interesting to me as Cohen explained it, it was hardly life changing. What happened next was.

It was getting close to the end of the first semester and Cohen presented a couple of lectures on the theory of relativity. I was overwhelmed. The idea that the speed of light was the universal speed limit and that every observer in a state of uniform motion would observe the same speed was incredible to me. There was also the claim that clocks in motion appear to run slower when observed by someone at rest. This also seemed incredible to me. What was the man talking about? In the course of this he said — I think as a joke although I did not see it that way — that only ten, or maybe it was twelve, people in the world who understood the theory. Much later I learned that when the great British astronomer Arthur Eddington, who in the 1920's had written a noted monograph on relativity, was asked if only three people in the world understood it. He probably thought of Einstein first, and himself, and jocularly replied, "Who is the third?" In any event, with all the lunar audacity of a Harvard freshman I decided to become either the eleventh or thirteenth person to understand the theory. I had a plan, which was to find a book that explained it.

Not only did I not possess the remotest background — mathematical or otherwise — to understand the theory, I did not even know what it meant to understand something like relativity. In high school I had taken Spanish so I knew what was meant by understanding a passage in Spanish. You translated into English which presumably you "understood." If you came to a hard word you looked it up in a dictionary where it was "defined" in terms of words that you "understood." When it came to poetry what was meant by understanding it was to be able to describe its meaning in terms of prose. In geometry what was meant by understanding was the capacity to prove something from the axioms, perhaps with the aid of diagrams. But what does it mean to understand the theory of relativity or the quantum theory for that matter? What it does *not* mean is to be able to "explain" the theory in terms of the theory that preceded it. In fact, this process should be the reverse. Be all that as it may, I went off to Widener Library to find a book.

I thought that I might as well get a book by Einstein since he surely understood the theory. There were a couple and I picked one called *The Meaning of Relativity*. I liked the title. It was the possibly worst choice I could have made. It was based on a series of lectures that Einstein delivered to specialists in 1921 at Princeton, although it has had revised editions. It is very sophisticated and highly mathematical. Apart from the opening sentence, "The theory of relativity is intimately connected with the theory of space and time"[d] I understood essentially nothing. I might have given up but instead I went to Cohen. He might have dismissed me as just another foolish freshman, but he didn't. He told me that in the spring semester there would be a course on about the same level as his which would be devoted to modern physics. He told me that it would be taught by a man named Philipp Frank, who was actually a friend of Einstein's and had just written a biography of him. Cohen said that he would have no objections if I took that course along with his, so I signed up.

The course gathered once a week, I think it was Wednesday afternoons, for two hours. We met in the large lecture hall in the Jefferson physics laboratories. When I showed up for the first lecture the room was full. Later I learned that people from other universities in the area came to audit. I had no idea what to expect. What would Professor Frank look like? When I first saw him he seemed perfect for the part. He was a smallish man who walked with a bit of a limp. Any hair left was in kind of a halo around his head. His accent was not easy to place. Although he had been born in Vienna — in 1884 — he had lived in Prague from 1912, when he succeeded Einstein at the German University, until 1938. Then he came to the United States where Harvard cobbled together some kind of appointment. He knew all sorts of languages including Persian which he had studied in night school in Vienna. I envisioned his languages as being piled up on top of each other like the buried cities of Troy. When he spoke English there appeared the shards of the other languages. Professor Frank would lecture for an hour and then make what he called "a certain

---

[d]Princeton University Press, Princeton, New Jersey, 1950, p. 1.

interval." After the "interval" he would answer questions or go into things with more depth, "If you know a little of mathematics."

His course also began with the Greeks. We learned that the Greek astronomers were committed to the notion that the celestial objects were attached to spheres that rotated uniformly. But this did not fit the facts. Planets, for example, periodically appear to move backwards. To account for this and still to "save the appearances" the uniformly moving plants moved in spheres around spheres — epicycles. Cohen had discussed this but Professor Frank added a new insight. Kepler in the 17th century had replaced all of this with single elliptical orbits. Professor Frank noted that such a motion can always be reproduced as a series of uniform circular motions so the two models cannot be distinguished if all one is interested in is describing the observed motions. Other criteria enter such as simplicity or "elegance." When it came to Newton, Professor Frank explained the story of the falling apple. He said, "Imagine an apple on a tree that grows to be as long as here to the Moon. Then the Moon is an apple on the tree and it too must fall in the field of the Earth's gravity." Then we came to relativity.

Professor Frank explained the relativity principle which was first stated by Galileo; the laws of physics are the same for observers at rest and observers moving uniformly with respect to these observers. Indeed, this second group of observers can perfectly well claim that they are the ones who are at rest. He told us that Einstein from very early on believed that this must apply to all laws of physics but he realized that there was a contradiction. If he was allowed to move with speed of light he could ride on a light wave. Then it would no longer look like a wave, so he would know that he was moving at the speed of light, which violates the relativity principle. It took him many years from his having first thought of this paradox to resolving it. You cannot in relativity move with the speed of light. Professor Frank had a simple way of explaining the transformations that take you from one system of coordinates to another that is moving uniformly. I have used it often when teaching the subject. He also told a funny story. He visited Einstein in Berlin a few years after Einstein had published his popular book on relativity. By this time he had married his second wife and was living with his two step daughters, one of whom was present. Einstein was explaining how his book was so simple that even his step daughter could understand it. He then left the room and Professor Frank asked her if it was really true that she understood it. Yes she said she understood everything except what was a coordinate system.

Following relativity Professor Frank turned to light. He told us about the double slit experiment of Young and he noted that if Young had been able to observe them he would have seen wave-like manifestations even from a single slit. There would be a build-up of intensities in front of the slit, but then there would be lesser peaks at spaces to either side. The profile would look like that of a mountain with its satellites to either side. All of this showed that in these experiments light behaved like a wave. But then Professor Frank told us about one of Einstein's 1905 papers — actually the one for which he won the Nobel Prize — in which he argued that in other experimental arrangements light would behave like a particle. In particular if you shined light on a metallic surface, electrons would be liberated from the metal — the photo-electric effect. This would also be predicted from classical physics. But what Einstein was proposing was that the energy of these liberated electrons did not depend on the intensity of the incident light, but only on its color. The more violet the light, the more energetic were the liberated electrons. In these experiments light behaved like particles (quanta) whose energy was proportional to the frequency of the radiation. Professor Frank said that if the intensity of the light was diminished so that it consisted of a single quantum, this quantum could still supply the energy to liberate an electron. What is light? It is a wave or a particle? It is both and neither. We then turned to electrons.

Professor Frank told us about the proposal of Louis de Broglie that electrons exhibited under certain conditions a wave character. He told us about the experiments — analogous to those of Young — in which electrons were made to impinge on a grating of slits. They were then deposited on a detector and one could observe the same kind of lines of intensity that Young had observed for light. What were electrons — waves, particles, both, or neither? Electrons, Professor Frank said, are electrons. But then he suggested a thought experiment — one that he said that had never been carried out —

although it has now.[e] Suppose you had a double slit and that your beam of electrons was so weak that only one electron at a time went through the system. What would happen? If both slits were open, Professor Frank told us, then over the course of time — at least according to the quantum theory — the single electrons would re-recreate the whole pattern. The pattern would not be, as one would expect for particles, the "sum" of the two open slit patterns. It would show the whole effect of the interference. There was no way of predicting where a given electron would land but you could predict using the theory where it was most likely to land. This to me was already amazing, but what he said next was more amazing. (It ultimately convinced me to go into physics.) Suppose you closed one of the slits and allowed the electrons to enter the apparatus one after the other. What would happen? The two-slit pattern would disappear and be replaced over the course of time by the one-slit pattern. The question that immediately occurred to me was how does the single electron going though the slit "know" that the other slit has been closed? Where does that information come from? By now I know that quantum mechanics does not provide an answer. In fact, it is not even the right question to ask, but at the time I didn't know this and was completely baffled and became determined to learn more and so I began a long course of study for which I eventually ended up with a Ph.D. in physics.[f]

I have often wondered if young people who are now first exposed to the theory have the same sense of wonderment that I did. After all, when I first learned it, the theory was only about twenty years old. All of the creators were still around lecturing and writing. Now the theory is nearly a century old. Do young people simply accept the rules and get on with it? What I do know is that some non-physicists do have this sense of wonder — even if, or mainly if, they do not really understand the theory. But they have absorbed enough

---

[e]See for example http://en.wikipedia.org/wiki/Double-slit_experiment for a review with many additional references.

[f]A reader who wants to know the details of this improbable saga can consult my autobiographical memoir *The Life it Brings*, Penguin, New York, 1988.

so that it has entered their creative imagination. A very interesting example of the genre is the playwright Tom Stoppard.

In the interests of full disclosure once again, I have to confess that I am a Stoppardian of the deepest dye. Plays like *Jumpers* or *Travesties* are for me intellectual champagne. The mixture of ideas and wordplay are for me irresistible. In the fall of 1994 I received a phone call from a woman named Anne Cattaneo. She identified herself as the editor of a journal named *The New Theater Review* — now called *The Lincoln Center Theater Review*. She told me that the Lincoln Center was mounting a play by Stoppard called "Hapgood" which had premiered in London in 1988. She also told me that the play involved quantum mechanics in some way. I had read about such a play and had been intrigued. Stoppard and quantum mechanics — what a pairing. Then she went on to tell me that *The New Theater Review* was planning an issue devoted to the play and its scientific implications. Would I like to contribute? She said that Stoppard was going to be one of the contributors. The prospect of sharing the stage with Stoppard was irresistible, so I agreed. The first thing that I did was to buy a copy of the play.

"Hapgood" is both the name of the play and the principle character. Her first name, which is almost never used in the play, is Elizabeth — Lilya in Russian. She is a spy, indeed the head of a coven of spies who refer to her as "mother." They are British except for one, Joseph Kerner, who is Russian. It was Hapgood who turned him but, in the course of things, the turning got a bit out of hand. Hapgood became pregnant and gave birth to a son, Joe. He is an adolescent and does not know who his father is. He is much more interested in playing rugby. Some of the play takes place during his school games. Kerner works at CERN, the elementary particle laboratory in Geneva. He is working on an anti-ballistic missile warhead that involves anti-matter. This is only marginally more absurd than other anti-ballistic missile programs I have heard about. There is a convoluted plot involving Kerner giving some or all of this information to the Russians. Periodically Kerner gives small lectures about the quantum theory and its relevance to human behavior. Schrödinger had enough trouble applying quantum theory to cats let alone human beings. I will shortly deconstruct one of these lectures, but first let me say something of how Stoppard appears to have gotten into this. What I know is what he wrote in the *The New Theater Review* an article which he entitled "The Matter of Metaphor." My article, which followed his, was called "A Trick of Light" and covered some of the same ground — although in considerably less detail — than what I have been discussing here. This is Stoppard,

"... Science for non-scientists is a boom area in publishing, and *Hapgood* is itself a fruit of a dozen books about quantum physics written for a general readership. If there were ever a general reader, I am one. I am not even a closet scientist, and certainly not a frustrated scientist. In fact, I probably owe it to my general lack of scientific education that the central image of *Hapgood* — the dual nature of light — excited me so much when I finally caught up with what is wearily familiar to anyone who stuck with physics past high school."[g] "Wearily familiar" does not describe my experience as I have tried to explain. Stoppard goes on,

"But, of course, it excited me for its potential as a metaphor. *Hapgood* is not "about" physics, it's about dualities. No — let the playwright correct the critic in me — *Hapgood* is not about dualities, of course, it's about a woman called Hapgood and what happened to her between Wednesday morning and Saturday afternoon 1989 just before the Berlin Wall was breached."

Stoppard continues,

"As a matter of fact, the story has much more to do with espionage than physics, but I won't deflect any compliments that might be going for a play with a reasonably plausible physicist on board, because the springs of the play are indeed science; it was only in looking around for a real world metaphor, that I hit upon the (John) le Carré world of agents and double agents. The physics came first, the woman called Hapgood came second."

He adds, "Incidentally, while several of the physicists who saw the play went out of their way to be kind about its general veracity, John le Carré, whom I came to know and, I seem to recall, who went on my tickets, managed to avoid mentioning the play from

---

[g] *The New Theater Review*, op. cit. p. 4.

that moment on; which I immediately recognized as the height of courtesy..."[h]

Stoppard does not tell us which physicists vouched for the play's "general veracity" or whether they both saw it and read it. Things go by pretty quickly in an actual performance. Nor does he tell us which popular books he read. It would be interesting to know. We do know, since he quotes from it, that he looked at Feynman's lectures.[i] This is a bit like my consulting Einstein's Princeton lectures to learn about relativity. Feynman gave these lectures to undergraduates at the California Institute of Technology (Caltech). Being Feynman, they are full of original insights. They are highly technical. I was told that the undergraduates found them pretty tough going although the faculty members, who audited them in droves, loved them. It is the last place I would send a non-scientist who wanted to get a flavor of the quantum theory. I was also told that Stoppard visited Caltech and actually sat for awhile in Feynman's empty office — for osmosis. What I want to do is to deconstruct one of Kerner's mini-lectures on the quantum theory. I do this in the spirit of the seminars that Bohr used to moderate where his many interruptions of the speaker were always accompanied by his saying that he was not there to criticize, but only to learn.

Kerner is talking to Paul Blair, another agent, who has informed Kerner that his cover as a double agent has been "blown." Kerner's career is over "except as a scientist." Kerner makes the point that a double agent is not like a giraffe — something definite but "more like a trick of the light." Then he explains,

KERNER: "Look. (*He points.*) Look at the edge of the shadow. It is straight like the edge of the wall that makes it. This means that light is particles: little bullets. Bullets go straight. They cannot bend round the wall a little, like the water round in a stone in the river."

I am surprised that none of the physicists Stoppard had contact with did not inform him that this characterization of light is entirely wrong. It had been observed as early as the seventeenth century that

---

[h] *The New Theater Review*, op. cit. p. 4.
[i] Feynman, op. cit.

shadows cast by light were fuzzy. Hooke and Huygens argued that this implied that light had a wave nature which manifested itself in this kind of observation. Light waves can interfere with each other whereas particles, as we usually conceive of them, cannot. When Einstein wrote his 1905 paper which claimed that under some circumstances light could exhibit a particle behavior, he began it by explaining how such a thing was possible given all the evidence that it was a wave. Thus began the idea of the complementary nature of light.

Blair responds,

BLAIR: (*Irritated*) "Yes. Absolutely."

KERNER: "So that's what. When you shine light through a gap in the wall it's particles. Unfortunately, when you shine light through *two* little gaps, side by side, you don't get particle pattern like for bullets, you get wave pattern like for water. The two beams of light mix together and..."

In what I have written above, I have explained that if you do the one slit experiment for light, or electrons for that matter, you get a pattern of light and dark places on the detecting screen. This happens because different parts of the slit produce waves that interfere. Even in the one slit case, the sort of question that bothered me as a freshman arises. Suppose you let the electrons through one at a time. Each electron will land at some spot on the detector, but as they accumulate they reproduce collectively the pattern of light and dark lines. How does the single electron know how to do this? How does it know where to go to play its role in the final pattern?

Conventional quantum mechanics has an answer, but it is not one that would have satisfied me as a freshman. By now I am used to it. The equation of Schrödinger has a solution which is a function of space and time. It you take this solution and square it, this gives the probability of finding the electron, say, at some place at a given time.[j] Where the function is large the probability of finding the electron is greatest. Nothing in the theory tells you where a given electron will

---

[j]Technically, the wave function is a so-called complex number so you must take the square of what is called the absolute value.

land, only where it is likely to land. This is one of the things that troubled Einstein. It also troubled David Bohm who produced his alternate interpretation of the quantum theory. In this there is a Schrödinger equation with a new term — the so-called "quantum mechanical potential" — whose solution acts as a "guidewave" for a particle such as an electron. The answer to the question of how does an electron know what to do in Bohmian mechanics is that the guide wave tells it. Back to the play.

BLAIR: "Joseph, I want to know if you're ours or theirs, that's all."

KERNER: "I'm telling you but you're not listening. Now we come to the exciting part. We will watch the bullets to see how they make waves. This is not difficult, the apparatus is simple. So we look carefully, and we see the bullets, one at a time. Some go through one gap and some go through the other gap. No problem. Now we come to my favourite bit. The wave packet has disappeared. It has become a particle again."

In 1927 there was a conference of leading physicists in Como, Italy to celebrate the centenary of the death of Alessando Volta. Quantum mechanics was something like two years old. Heisenberg had just published his uncertainty principle. Both Bohr and Einstein were at the conference. Bohr introduced his notion of complementarity. Einstein had a little surprise for him. It was a thought experiment involving double slits. But these were mounted on rollers or wheels which made the apparatus movable. Now suppose you take a position on the detector above both slits. An electron comes to the lower slit, say. To reach your position it must change its direction which means that it must change its momentum. Since momentum is conserved the apparatus acquires this momentum in reverse and that can be detected since it will recoil on the wheels. This momentum can be measured. If the electron comes through the upper slit, there will also be a momentum change but less since the angle it goes through is smaller. Hence by measuring the momentum recoil of the apparatus we can tell through which slit it went, so it would appear that in one and the same experimental arrangement we can determine both the particle and wave aspects of the electron. The wave aspect would be manifest by the intensity patterns. But Bohr had an answer. The momentum of the apparatus must be measured to a certain accuracy in order to make this determination. But by Heisenberg's uncertainty principle this limits the accuracy of the position of the slits. If you work out the details this inaccuracy is greater than the distance between the first maxima in the interference pattern. The slits are effectively as wide as the interference pattern so there is no interference pattern and Einstein's thought experiment does not work. Sometimes one reads in popular books that this set up shows that the electron can be in two places at once. The electron is in a "place" only when you observe it on say the detecting screen. If someone tells you that it is also in another place the proper answer is "show me." Recall Kerner's remark I quoted at the beginning, "An electron can be here or there at the same moment." In so far as it is here or there it is either one or the other. The play goes on.

BLAIR: "How?"

KERNER: "Nobody knows. Somehow light is continuous and also discontinuous. The experimenter makes the choice. You get what you interrogate for. . ."[k]

Blair's question "How" is exactly the kind of question I asked as a freshman and to which there is no answer at least in the quantum theory. Stoppard's quantum theory reminds me of a conversation I had with Marvin Minsky when we were both undergraduates. He later went on to become one of the founders of artificial intelligence. He showed me a drawing he had made of a bird. "This looks like a bird," he said, "but no bird looks like this."

Professor Frank died the 21st of July 1966. Later that year there was a memorial for him at Harvard at which I was one of the speakers. His widow Hania whom he had married in Prague when she was one of his students was there. She belonged to an intellectual circle which included Kafka. She was a remarkable and very lively woman. I have an ineluctable memory of calling one night to speak to Professor Frank. Hania answered and in her inimitable accent said, "We are here singing English folk songs and Philipp has gone away." I wish that I could have discussed Stoppard's play with him.

---

[k]Hapgood, op. cit. pp. 500–501.

4. Dirac, Some Strangeness in the Proportion

Reprinted with permission from American Journal of Physics
77, 979 (2009). Copyright American Association of Physics
Teachers

> "There is no exquisite beauty without some strangeness in the
> proportion...."
>
> Francis Bacon

If like myself you have spent a lifetime among theoretical physicists
and mathematicians you will have become accustomed to eccentric
behavior. There are of course degrees from the innocuous to the
truly pathological. Some examples. When I joined the Institute for
Advanced Study as a temporary member in the fall of 1957, there
was a French mathematician who had decided to live his life in a
unit of 27 hour days. Why he did, I have no idea. That he did,
since like many mathematicians he worked by night and slept by
day, went largely unnoticed. Norbert Wiener was a distinguished
mathematician who spent his career at MIT. In order not to lose time
he would read as he walked its corridors, one hand holding a book and
the other acting as a guide along the walls. At seminars he sat in the
front row apparently asleep only rousing himself when his name was
mentioned or when he thought it should have been mentioned. These
were benign eccentricities which I think were consciously adopted to
add to the legend of the persons involved. The case of the great

logician Kurt Gödel is quite different. His eccentricities merged on serious mental illness. Indeed in the 1930's he spent several months in mental institutions. Ultimately he starved himself to death.

During the years 1944–1948 Einstein had as his assistant the mathematician Ernst Gabor Straus. Around the centennial of Einstein's birth (1979) there was a meeting in Princeton, which I attended. Straus gave a delightful reminiscent talk. Some of it was devoted to Einstein's friendship with Gödel. This unlikely pair would walk each other home. There are photographs showing Einstein in his informal dress and Gödel looking more like a professor, which he was at the Institute. In 1948, he became an American citizen, and not without an obstacle. Einstein accompanied him to Trenton for the swearing in. Gödel informed the judge that he had found a logical inconsistency in the Constitution which was making it difficult for him to pledge the necessary allegiances. The judge told him, more or less, to shut up and swear. A few years later Einstein came into Straus's office and said, "Now Gödel has done something really crazy." Straus could not imagine what such a thing could possibly be. "He voted for Eisenhower," Einstein explained. I met Gödel only once. Early in the fall term of my first year at the Institute, Robert Oppenheimer put on a dance in the Institute cafeteria for the members. We are all standing on the edge of what had been turned into a dance floor when I spotted Oppenheimer, his wife and Gödel. What could have persuaded Gödel, who was generally as elusive as a snail darter, to agree to come to the dance, I cannot imagine. But there he was and Oppenheimer was going around the room introducing him. I was presented as Doctor Bernstein and Gödel said, "I knew your father in Vienna." I explained that my father was in Rochester, New York and had never set foot in Vienna when Gödel was there. "I knew your father in Vienna," Gödel repeated. I let the matter rest and he moved on.

Around 1693, Isaac Newton had a serious nervous breakdown. The reasons for this have been the subject of much speculation. One theory is that it was mercury poisoning brought on by his alchemical experiments. There are authors who claim he was autistic, a modern retrospective diagnosis that covers a large class of behavioral disorders. A clue may be in a very strange letter he wrote to the philosopher John Locke with whom he had had some theological disagreements but nothing that would produce such a letter. What is especially odd about it is that it was written after Newton had a spontaneous return to normalcy. Locke was quite unaware that Newton had had this breakdown. One wonders what he thought of the letter. It seems Newton never had a recurrence. However, he did no more serious science and had an agreeable old age in London where he became Master of the Mint. One of the perks of the job was the arresting and subsequent drawing and quartering of "clippers and coiners" — counterfeiters. Newton took great satisfaction from his role. Here is the letter. The spelling is in the original.

"Sir, being of the opinion that you endeavoured to embroil me with women and by other means, I was so much affected with it as that when one told me you were sickly and would not live, I answered 'twere better if you were dead. I desire you to forgive me this uncharitableness."

Niels Bohr, who knew very many of the physicists of the twentieth century, said that the "strangest man"[a] who ever spent time at his institute in Copenhagen was the great English theoretical physicist Paul Adrien Maurice Dirac. *The Strangest Man* is the title of a very moving biography of Dirac by Graham Farmelo. It would have been easy to simply fill a biography like this with Dirac stories of which there is a cornucopia — many of which are actually true. But Farmelo does much more than this. He has met and spoken with people who knew Dirac including the surviving members of his family. He has been to where Dirac lived and worked and he understands the physics. What has emerged is a five hundred and fifty eight page biography which is a model of the genre. Dirac was so private and emotionally self-contained that one wonders if anyone really knew him. Farmelo's book is as close as we are likely to come.

Dirac was born the 8th of August 1902 in Bristol, England. Unlike Archibald Leach — Cary Grant — born two years later, he never lost his Bristol accent. It is unlikely that Dirac ever saw a Cary Grant

---

[a]This is quoted in *The Strangest Man*, by Graham Farmelo, Faber and Faber, London, 2009, p. 120. Farmelo gets this anecdote from Kurt Gottfried, arXiv:quant-ph/0302041v1-5 (Feb 2003). Gottfried heard it directly from Bohr.

movie. He was very fond of Mickey Mouse and sat through three successive performances of Stanley Kubrick's "2001." This would have pleased Kubrick who was something of a physics groupie. Dirac's father was Swiss and after an itinerant career teaching languages on the continent he had gotten the job of Head of Modern Languages at the Merchants Venturers' School in Bristol. He met Dirac's considerably younger mother Florence on a visit to the library where she worked. They were married in 1879 and Dirac's older brother Felix was born a year later. A sister Betty was born in 1906.

As a teacher Dirac's father had the reputation of being fair but strict. At home, if Dirac can be credited, he was an abomination. Dirac rarely talked about his childhood but when he did he told of meals where the other two children and his mother ate in the kitchen while Dirac and his father ate alone. Dirac's father insisted that Dirac speak French at the table and since he could not do this very well he remained silent. He was forced to finish all his food even though he was full and the remaining food made him sick. Dirac explained — he did so to me the one time I interviewed him for a *New Yorker* profile which was never written — that this childhood experience is what made him so silent throughout his life. On the other hand, the letters that Farmelo quotes to Dirac from his father are affectionate. He was very proud of what his son became. Dirac's only real childhood companion was his older brother Felix. The marriage became more and more unhappy over time. Dirac's parents never divorced although, while they were both alive, his mother sent Dirac a stream of complaining letters which Dirac simply ignored. Once he left his parent's home he went back as little as possible.

Both Dirac and his brother attended the very nearby Bishop Road School. There is nothing in the record that Farmelo can find that indicates that Dirac was considered intellectually exceptional in any way, and there is nothing that reflects a disturbed home life. The one way the two brothers stood out is that they could both speak French. From the Bishop Road School the boys matriculated to the Merchants Venturers' School where their father taught. It was here that Dirac began to stand out. He rapidly became the top of his class. The school was widely known for its technical education — the training of future engineers for example. One of his teachers thought that Dirac was so exceptional that he almost became his private tutor. Meanwhile his older brother was compiling an ordinary record and was a lesser parental favorite to boot.

Housed on the same premises was the Merchant Venturers' College which was basically an engineering high school. Felix was the first to enroll followed by his brother. Felix graduated first but only with third class honors. It was only too painfully clear that as far as academics were concerned the two brothers were on separate planets. It appears as if Dirac was unable, or unwilling to reach out to his brother who soon left home. As he was approaching graduation his father suggested that he apply to Cambridge. Whatever loathing Dirac manifested towards his father it should never be forgotten that it was he who urged Dirac to take the next step. If he had not done that, Dirac might have ended up as an engineer and the history of twentieth century physics would have looked quite different.

In 1921 Dirac's father wrote to St. John's college in Cambridge — he applied to the colleges and not the university — to see if his son could apply for a scholarship. Dirac won one but it paid too little for him to make use of it. He then looked for a job. There was a depression and he got none. This turned out to be very fortunate because Roland Hassé, who was chairman of the mathematics department at Bristol University, invited Dirac to take courses there free of charge. Dirac took full advantage and acquired a solid grounding in mathematics and physics. He took a mathematics degree from Bristol, but his father still wanted him to attend Cambridge. With Hassé's help a fellowship to St. John's was arranged. But there was a requirement that the student pay five pounds in advance — five pounds which Dirac did not have. His father gave him the money. Late in life Dirac learned that it had not actually, as he had claimed, been his father's money and found one more reason to detest him which, under the circumstances, seems a little much. Incidentally Hassé wrote a letter of recommendation to Cambridge which is a kind of classic. It reads in part, "He is a bit uncouth, and wants some sitting on hard, is rather a recluse, plays no games, is very badly off financially."[b]

---

[b]Farmelo, p. 53.

There is a Russian proverb that says, "To live life is not like crossing a field." There were so many ways that Dirac might not have become a physicist and certainly might not have gotten to Cambridge, but in 1923 he did. He was fortunate in that his supervisor Ralph Fowler was a well-known mathematical physicist whose eye was always out for exceptional students. It was clear from the beginning that Dirac was one. Dirac also learned to be in the company of others especially at meals although he said little or nothing — his colleagues in Cambridge jokingly defined a unit of a "dirac," which was one word per hour.[c] He did make a few friends, especially the Russian physicist Peter Kapitza who was working in the Cavendish Laboratory with Ernest Rutherford, the discoverer of the atomic nucleus, who was the leading physicist at the university. Kapitza was the mirror opposite of Dirac — sociable to the extreme — but their friendship lasted until Kapitza's death in 1984. Kapitza was well known for his work on the physics at low temperatures for which he won the Nobel Prize in 1978. It was from Fowler's lectures that Dirac first learned about the quantum theory — the "old" quantum theory.

The old quantum theory began with work of people like Max Planck and Einstein. It was an uneasy mixture of classical and quantum ideas. Nowhere was this more clear than in Bohr's theory of the atomic electrons. After the nucleus was discovered it became clear that the electrons had to be in orbits outside it. There were several problems with this picture. Classical physics predicted that these orbiting electrons would radiate. But when they did so they would lose energy and fall into the nucleus making matter unstable. Moreover the frequencies of this radiation would in no way resemble the beautiful spectral patterns that these atoms emit and which identify them like fingerprints. In 1913 Bohr proposed a new model for these orbits. The energies of the orbits were "quantized" — their magnitudes were limited to certain values. When an electron jumped from a higher to a lower energy orbit it emitted radiation whose energy was given by the difference in energies associated with the two orbits. The orbit of least energy — the "ground state" — was absolutely

---

stable. This model was elaborated on for the next two decades. In May of 1925 Werner Heisenberg, who was then an assistant to the German physicist Max Born in Göttingen, came down with a severe case of hay fever. To seek relief he went to the more or less treeless island of Helgoland in the North Sea. It was here that he had the inspiration that lead to modern quantum mechanics. When I interviewed Dirac he was insistent that the credit for inventing quantum mechanics was due to Heisenberg. Heisenberg was a year older than Dirac. They were both in their early twenties at the time.

Heisenberg's guiding principle was his insistence that physical theories should only involve quantities that are observable in principle. He later claimed that this was something he learned from Einstein. Einstein noted that he might have said something like this when he was younger but that he no longer believed it. In any event, Heisenberg noted that the orbits of electrons around nuclei were not observable. When, later on, he invented his uncertainty principles he argued that if you tried to observe an electron in its orbit it would require a light quantum that was so energetic that it would knock the electron out of the atom destroying the orbit. What was observable was the radiation coming from the atoms. The intensity of this radiation, which was also observable, was related to the probability of the electron making a jump from one level to another. Heisenberg wrote down expressions for these probability amplitudes and noticed that they obeyed very strange algebraic relations.[d] In the algebra we first learn in school if we have quantities x and y then $xy = yx$. The term of art for this is that x and y "commute." With Heisenberg's amplitudes A and B, AB was not equal to BA. There was a very definite multiplication law dictated by the physics. With this in hand he went back to Göttingen to show Born what he had done, thinking that Born might find it total nonsense. Born took one look and immediately understood what Heisenberg's mathematics was.

---

[d]The amplitudes depend on which energy levels they refer to. If a transition is from level characterized by the integer n to one characterized by an integer m then the amplitude A_{nm} is a function of these levels. It is the multiplication of these amplitudes that does not commute.

In the 19th century mathematicians had invented something called "matrix algebra." These matrices were arrays of numbers that multiplied in the same way that Heisenberg's expressions were multiplying. Without knowing it Heisenberg had introduced matrices into the quantum theory. Heisenberg wrote up a paper and gave it to Born to decide whether it was worth publishing. Then he headed to England where he had been invited to speak at Cambridge. Born submitted the paper and began working on its generalizations with a brilliant assistant named Pascual Jordan. Jordan had his own psychological problems. He had a stutter so thick that he could hardly talk. He later became a very enthusiastic Nazi.

Heisenberg's paper is notoriously difficult to read.[e] He was inventing new physics using mathematical methods he did not fully understand. I doubt that anyone not interested in the history of the subject would try to read it. One the other hand, the paper of Born and Jordan which was published soon after, was a model of clarity. You could teach part of a course in the quantum theory using it. It begins with a pedagogical discussion of matrices — something that an undergraduate physics major would now be familiar with. Then there are the applications to quantum physics. I would like to call your attention to two. In quantum theory the position x of a particle is represented by a matrix — these matrices are usually referred to as "operators". Likewise the momentum p is also represented by an operator. If we form the product xp it is not equal to px. Born and Jordan show that xp-px is proportional to the constant h that Planck first introduced to characterize the quantum nature of radiation. This constant is very small which is why we are not aware of quantum effects in our daily lives. The expression xp-px is so important in the quantum theory that it is referred to as the "canonical commutation relation." Born and Jordan also showed how x and p evolve in time under the influence of forces. The equations they derived are usually called the Heisenberg equations although they are nowhere to be found in Heisenberg's paper. Heisenberg was awarded the 1932

---

[e]English translations of this paper and that of Born and Jordan as well as the paper of Dirac can be found in *Sources of Quantum Mechanics*, edited by B.L. van der Waerden, Dover Publications, New York, 1968.

Nobel Prize for his work. He felt so strongly that Born should also have gotten a prize that he went to Switzerland to post an uncensored apologetic letter to Born who had left Germany. Born was finally awarded his Nobel Prize in 1954.

On the 28th of July 1925, Heisenberg gave a lecture on his new work at the "Kapitza Club" in Cambridge, an informal society that Kapitza had organized where physicists could discuss their work. It is not clear whether Dirac was at the lecture. The previous March there had been a family disaster. His older brother Felix had committed suicide. Dirac's reaction, which he described forty years later, almost defies understanding. He remarked that "My parents were terribly distressed. I didn't know they cared so much [...] I never knew that parents ought to care for their children, but from then on I knew."[f] Dirac returned to Bristol to spend part of the summer with his family. In September he received an envelope from his tutor Ralph Fowler. It contained the proofs of Heisenberg's paper with a note that read "What do you think of this? I shall be glad to hear."

When Dirac studied the paper he decided that Heisenberg's multiplication laws were the key element. He too did not seem to know about matrices but he realized that he had seen something like these multiplication laws before. There is a way of formulating classical mechanics using what are called "Poisson brackets." These brackets have algebraic properties which are similar to what Dirac called "Heisenberg products," or so he remembered. This happened on a Sunday night sometime after he had returned to Cambridge and none of the libraries were open. The next day he was able to verify his recollection and he made the assumption that the Heisenberg products and the classical Poisson brackets were proportional to each other and that the constant of proportionality was again Planck's constant.[g] From this assumption the canonical commutation relations and the Heisenberg equations emerged. The

---

[f]Farmelo, p. 79.
[g]Suppose A and B are dynamical variables that are functions of the position q and the momentum p — for simplicity I will take these to be one dimensional the Dirac assumed that $AB - BA = i\hbar(\partial A/\partial q \partial B \partial p - \partial B/\partial q \partial A/\partial p)$. The quantity in the parens is the Poisson bracket of A and B. If you

paper of Born and Jordan had not yet come to Cambridge so this was an original discovery with a new insight. Fowler was very impressed and moved to have Dirac's paper published in the Proceedings of the Royal Society under its title *The Fundamental Equations of Quantum Mechanics*, with great haste so as to assure Dirac's priority. For Dirac, this was just the beginning. The paper became the basis of his Ph.D. thesis.

Apart from Fowler there was not much, if any, interest shown by Cambridge people in Dirac's paper which was sent in pre-publication form to Heisenberg. Heisenberg noted that some of the results were in Born and Jordan, but Dirac's insights were new. Born and Jordan's paper had not yet come to Cambridge and Dirac made no reference to it himself. In the meanwhile an entirely new development in quantum theory came unexpectedly from Erwin Schrödinger. One says "unexpectedly" because Schrödinger, who was a generation older than either Dirac or Heisenberg, had not done anything significant in atomic physics and was prepared to spend the rest of his life teaching physics and philosophy in a provincial Austrian university until 1918 when Austria lost the province where he was teaching. He then went to Zurich to resume his career in theoretical physics. In 1925 he had an epiphany which eventually immortalized him. The year before, the French doctoral candidate Louis de Broglie — who was in his early thirties — proposed that particles like electrons also had a wave nature. This complemented Einstein's idea which he introduced in 1905 that light had a dual wave and particle nature. But waves propagate in space and time and satisfy an equation of propagation. Schrödinger proposed an equation — it now bears his name — for this wave propagation. Schrödinger's "wave mechanics" appeared to be a new and different version of the quantum theory. It turned out that matrix mechanics and wave mechanics were two different representations of the same theory so that either one could be used to solve problems, and this was shown by both Schrödinger and Dirac.

In 1926 Dirac began a year-long travelling fellowship which he split between Bohr's institute in Copenhagen and Max Born's group

---

take A = q and p = B the canonical commutator follows. But notice that this is an assumption that cannot be derived from classical physics.

in Göttingen. He first went to Copenhagen. This is where he made
his impression on Bohr. Bohr was almost not capable of simply writing a paper. A young assistant would be drafted to take down end-
less rounds of constantly modified dictation. It was Dirac's turn but
after a few minutes he said to Bohr, "I was always taught not to
start a sentence until I knew how to finish it."[h] That was the end of
that. Dirac kept largely to himself although, surprisingly, he greatly
enjoyed spending time with Bohr's family. It was so different from
his own family with Bohr's rambunctious boys and their very affectionate parents. One wonders if he was ever able to communicate his
feelings to Bohr. It was in Copenhagen where he first met Heisenberg and Wolfgang Pauli. He developed a friendship with Heisenberg
which lasted all of Heisenberg's life and which endured despite the
fact that they were on opposite sides during the Second World War.
He did not much like Pauli who was notorious for biting *ad hominem*
criticisms of people's work. It was at Göttingen where Dirac first met
Oppenheimer.

They might have met in England where Oppenheimer had a pretty
miserable time. He and Dirac had overlapped in Cambridge but while
Oppenheimer was nursing his neuroses Dirac was busy formulating
the quantum theory. The two of them were about the same age —
Dirac being a little older. In Göttingen they boarded with the Carios
family in a rather lavish villa which had fallen on somewhat hard
times. The two men became friends and when Oppenheimer became
director of the Institute for Advanced Study Dirac gave a standing
invitation. That is where I first saw him in the fall of 1958. We were
in the midst of a seminar and the door opened and there was Dirac
wearing what looked like high rubber boots. It turned out that he was
clearing some sort of trail in the woods behind the Institute. There
is a photograph of him from that period in Farmelo's book. He is
carrying an axe. Dirac must have learned, when they first met, that
Oppenheimer had a great interest in poetry. The two men used to
walk on the walls surrounding Göttingen and on one occasion Dirac
remarked, "I don't see how you can work on physics and write poetry

---

[h]Farmelo, p. 111.

at the same time. In science you want to say something nobody knew before in words everyone can understand. Whereas in poetry..." Farmelo has this anecdote but he completes the sentence. When I heard Oppenheimer tell it he let the listener complete it.

Oppenheimer wrote a very respectable thesis with Born on the quantum theory of molecules — an important application that is still used — but it must have been odd to be in the presence of someone about his own age who was actually one of the creators of the theory. At the time Dirac was developing a quantum theory of radiation — quantum electrodynamics. He gave Oppenheimer a draft of what he had been working on and Oppenheimer immediately recognized its importance. Dirac had also discovered some of the problems — the infinities that plague the theory. After the Second World War a practical way of dealing with these infinities was found but Dirac thought that it was ugly and was never satisfied with the theory. In 1927, Dirac was back in Cambridge and the work that he did that year, had he done nothing else, would have labeled him as one of the greatest physicists of the twentieth century.

The problem was how to find an equation for the wave function of the electron that was consistent with Einstein's theory of relativity. There was a prescription as to how to go about finding such wave equations. You first write down an expression for the energy of the particle in terms of its momentum, mass and possible forces acting on it. The simplest case to start with is when no force is acting. Then you replace the momentum by its quantum mechanical expression and out of this you get the quantum mechanical wave equation. The problem in relativity is that the simple expression is for the square of the energy. It is in essence given by the sum of the squares of the momentum and the mass, that is $E^2 = p^2c^2 + m^2c^4$, where c is the speed of light. Indeed before Schrödinger found his equation he did just this which led to an equation that describes some particles but not the electron.

The problem is that the electron has another property in addition to its mass and momentum. It has what it known as "spin." Sometimes one finds this explained with a picture of an electron spinning like a top. But a top can spin in any direction and an electron can spin only in two. This is a purely quantum mechanical fact. Pauli had shown how to construct matrices that represent spin and it was straightforward to add these spin terms to the Schrödinger equation. But not only was this *ad hoc* but the equation itself was only valid for electrons that move slowly compared to the speed of light. What Dirac did was to find an equation that resolves both of these difficulties at once. It is one of the most beautiful equations in all of physics.

Schrödinger had derived his relativistic equation by considering the square of the energy. What Dirac decided to do was to consider the energy itself. This meant he had to look for the square root of Schrödinger's equation. What Dirac did was to factor the expression $E^2 - m^2c^4 - p^2c^2 = 0$, into two equations that were linear in the operators for E and p. To make the product of these linear equations equal to the original expression Dirac found that the coefficients of these operators were not ordinary numbers but rather matrices whose commutation relations were determined by this condition of equality. The simplest representation of these matrices was in four dimensions. The Dirac equation is an equation in four dimensions. There are four solutions for a given momentum and mass. Each of these solutions represents a particle with a direction of spin. But what particles do they represent? It is here that the difficulties presented themselves.

The Dirac equations had both positive and negative energy solutions. Perhaps this could have been anticipated since in some sense we are taking a square root to derive the equations. Clearly a negative energy particle is not physical. So long as we are talking about free particles, these negative energy solutions are not a problem. In solving the equations of physics we often run into undesirable solutions. They might have the wrong boundary conditions or might be singular or whatever. We simply throw out these "bad" solutions and go on our way. But electrons are by the nature of things coupled to electromagnetic fields. A free electron cannot without violating the conservation of energy make a quantum transition to a state with negative energy. But if it is coupled to the electromagnetic field it can. Radiation quanta can be emitted which make up the deficiency in the energy. Why then are all the electrons not simply disappearing?

To confront this dilemma Dirac invoked the Pauli principle. This was somewhat ironic because, to Pauli, Dirac's solution was an example of *verzweiflungsphysik* — desperation physics — and he said so in no uncertain terms. Dirac wanted to assume that there was a "sea" of these negative states electrons which was filled up by electrons. Since the Pauli principle prevents two electrons occupying the same state this transition to negative energy states would be forbidden. Dirac argued that there might be "holes" in this sea and that these holes would behave like positively charged particles with positive energy. The transition would then mean that a positively charged and a negatively charged particle would annihilate each other with the emission of radiation. The only positively charged particle known at the time was the proton so it was selected. Oppenheimer showed that this choice had disastrous consequences for the stability of matter. Moreover, the mathematician Herman Weyl showed that the theory requires that these two Dirac particles have the same mass. Thus the positive electron, the first example of anti-matter, was born.

In 1931 Dirac published a paper called "Quantized Singularities in the Electromagnetic Field"[i] The burden of the paper has to do with the highly speculative idea that single magnetic poles — monopoles — might exist. This is still something that is explored although none have been found. Almost as a throwaway there occur the following two paragraphs,

"A recent paper by the author may possibly be regarded as a small step according to this general scheme of advance. The mathematical formalism of that time involved a serious difficulty through its prediction of negative kinetic energy values for an electron. It was proposed to get over this difficulty, making use of Pauli's Exclusion Principle which does not allow more than one electron in any state, by saying that in the physical world almost all the negative-energy states are already occupied, so that our ordinary electrons of positive energy cannot fall into them. The question then arises to the physical interpretation of the negative-energy states, which on this view really exist. We should expect the uniformly filled distribution of negative-energy states to be completely unobservable to us, but an unoccupied one of these states, being somewhat exceptional, should

---

[i] *Proc. Roy. Soc. A*, Vol. 133, p. 60.

make its presence felt as a kind of hole. It was shown that one of these holes would appear to us as a particle with a positive energy and a positive charge and it was suggested that the particle should be identified with a proton. Subsequent investigations, however, have shown that this particle necessarily has the same mass as an electron and also that if it collides with an electron, the two will have a chance of annihilating one another much too great to be consistent with the known stability of matter."

Dirac goes on,

"It thus appears that we must abandon the identification of the holes with protons and must find some other interpretation for them. Following Oppenheimer, we can assume that in the world as we know it, *all*, and not nearly all, of the negative-energy states for electrons are occupied. A hole, if there was one, would be a new kind of particle, unknown to experimental physics, having the same mass and opposite charge to an electron. We may now call such a particle an anti-electron. We do not expect to find any of them in nature, on account of their rapid rate of recombination with electrons, but if they could be produced experimentally in high vacuum they would be quite stable and amenable to observation. An encounter between two hard γ-rays (of energy at least half a million volts) could lead to the creation simultaneously of an electron and anti-electron, the probability of these processes being of the same order of magnitude as that of the collision of two γ-rays on the assumption that they are spheres of the same size as classical electrons. This probability is negligible, however, with the intensities of γ-rays at present available."

And Dirac concludes,

"The protons on the above view are quite unconnected with electrons. Presumably the protons will have their own negative-energy states, all of which are normally occupied, an unoccupied one appearing as an anti-proton. Theory at present is quite unable to suggest a reason why there should be any differences between electrons and protons."

This is hardly a clarion call to experimenters especially as it appears almost a throwaway in the paper.

In the summer of 1932 a young experimenter named Carl Anderson at Caltech was using a cloud chamber to detect and photograph tracks from charged cosmic ray particles. He had heard of the Dirac equation in some lectures by Oppenheimer but had not associated it with the prediction of the existence of an anti-electron. He had put a magnetic field on his cloud chamber so he could tell the sign of the charge of a particle by how it was deflected. He could also tell whether or not a particle was an electron by the kind of tracks it made. Anderson's particles were cosmic rays coming from outer space. On the second of August he found the track of what looked like a positive electron. By the end of August he had found two more. He was not sure whether to publish on the basis of such a small number of events, but he was talked into doing it. This was fortunate for him because in 1936 he was awarded the Nobel Prize for the discovery of what by then was named the "positron." It was also fortunate for Dirac.

In 1933 he shared the Nobel Prize with Schrödinger. That same year Heisenberg picked up the prize that he had been awarded the year before. I think that if the positron had not been discovered when it was, Dirac would not have gotten the Nobel Prize that year. He had of course done wonderful work. But some of it had been done by others as well. To take an example, in 1926 Dirac had made an analysis of the statistics of particles like electrons that obey the Pauli Exclusion Principle. It was an important piece of work but it was also done by Enrico Fermi. It had also been done by Jordan but that was never published. These statistics are now known as "Fermi-Dirac" statistics. Dirac's quantum theory of radiation was pioneering but not many people understood it and it seemed to raise more questions than it resolved. But about the Dirac equation and its prediction of anti-matter, there could be no argument. Indeed, this work is what the Nobel committee cited. At first Dirac wanted to turn down the prize because of the publicity involved. But Rutherford explained to him that if he turned it down there would be even more publicity. He was, it seems, allowed to take one guest to Stockholm for the ceremony and he chose his mother.

In 1930, Dirac published the first edition of his monograph *The Principles of Quantum Mechanics*. I would not call it a "text." There are no diagrams or figures. There are no problem sets and almost no reference to experimental observations. However Einstein felt that it was the most logically perfect exposition of the theory he had ever found. And Dirac seems to have shared his view. In 1932, he became Lucasian professor at Cambridge — a post held by Isaac Newton and later held by Stephen Hawking. He taught a course in quantum mechanics. Some years later Freeman Dyson took the course. He said that Dirac simply read from his book. When questioned about this Dirac said that he had given a great deal of thought as to how to present the subject and had decided that what was in his book was the best way and that there was no need to change it. I own a copy of the third edition. When Dirac was at the Institute for Advanced Study in 1968 I asked him to autograph it. Under his printed name he wrote "P.A.M. Dirac" and that is all.

On the second of January 1937, Dirac got married. This astounded almost every one who knew him. He married the sister of the physicist Eugene Wigner. Her first name was Margit but everyone called her Manci. She was a widow with two children who lived in Budapest. Dirac had met her in the fall of 1934 when both of them were in Princeton. She and her brother were having lunch in a restaurant and when Dirac came in alone he was asked to join them. I think that it is fair to say that in the courtship that followed she played the more active role although Dirac did write some letters that could pass for being romantic. People said that he used to introduce her as "Wigner's sister" even after they were married. I met her only once. In the early 1960s I did a brief stint at the Boltzmann Institute in Vienna. Its director was the Austrian theoretical physicist Walter Thirring. Thirring actually arranged a visit with Schrödinger who was living in retirement in Vienna. I was in such awe that I didn't ask him the dozens of questions that have occurred to me since. Dirac and his wife had come to Vienna. Thirring asked me if I could take them around and show them the sights. They could not have been more charming or appreciative.

Dirac's relationship with his brother-in-law was I think a bit distant. Wigner was a Russo-phobe of the deepest dye, while Dirac had always had a warm feeling about the country and toward the Russian physicists that he knew. The late Valentine Telegdi, who was a physicist, told me that he witnessed the following exchange. Manci was out of town and on a Wednesday Wigner had come to the Institute to ask Dirac to have dinner that night. Dirac said no so Wigner tried for Thursday and Friday with the same response. Finally he tried for Saturday and Dirac agreed. Wigner wanted to know why Dirac had agreed to Saturday. "The Institute cafeteria is closed on Saturday," was the response.

Dirac was quoted during the years as saying a number of things that most people would find funny. I do not think that he ever said anything to be deliberately funny. His mind seemed to work in a very literal sense. The things that he said were literally true but the context was skewed. Here are two examples, one of which I witnessed myself and the other of which was told to me by Sir Rudolf Peierls who was good friend of Dirac's. The one I witnessed occurred at dinner in the Institute cafeteria. Dirac would frequently join us when his wife was out of town. One night I was discussing the financial fortunes of Hans Bethe with a colleague who was from Cornell where Bethe taught. I knew that Bethe was a very well-paid consultant and that he had a high salary. Moreover he lived very modestly. I asked my colleague what Bethe did with all his money. My colleague said he thought that Bethe invested it. I said that this did not solve the problem because then Bethe would have even more money. Suddenly Dirac who had been silent said, "Maybe he loses it." I am sure that he was not saying this to be funny but rather to simply call attention to a possibility that we had overlooked.

Peierls told me that when his daughter went to Cambridge he told her to look up the Diracs. She was duly invited to tea. Manci asked her if she knew people her own age at Cambridge and Miss Peierls said that she did not know many. Manci then asked her husband if he did not have a student or two that Miss Peierls might meet. Dirac thought and then answered, "I had a student once, but he died." Dirac did not really want students. The idea of his creating an institute like that of Bohr's or Born's is absurd. On one of the evenings at the Institute when Dirac joined us for dinner, we were discussing collaboration in science. One of my colleagues asked Dirac if he ever collaborated. "The really good ideas," he said, "are had by only one person."

The Diracs spent the war in England. At the beginning of it, Peierls and Otto Frisch made an estimate of how much uranium-235

would be needed to make a nuclear weapon — the "critical mass."
They came up with a relatively small number that alarmed them.
They assumed, incorrectly as it happened, that Heisenberg and the
other German physicists must have made the same calculation and
were actively trying to separate uranium-235 from the vastly more
common uranium-238. The Germans did consider methods of separation but Heisenberg apparently never made a calculation of the
critical mass during the war. In any event Frisch and Peierls alerted
scientists working for the British government and this led to the
founding of a nuclear weapons program. Dirac got involved. My guess
is that he was recruited by Peierls. He took it on himself to study the
theory of isotope separation involving for example centrifuges. When
people in this business refer to the "Dirac equation," they are not
talking about relativistic electrons. They are talking about the equation that Dirac derived for the maximum output per unit length of a
rotor with a given peripheral velocity that is possible in principle. In
doing this Dirac introduced concepts that have been in use eversince.
If you read his brief paper without knowing who wrote it, you would
very probably not think of Dirac. You would know that, whoever
wrote it must have been a very good physicist or engineer. Indeed, if
Dirac had not gone into physics, he might well have become one of
the best engineers who ever lived. (See the appendix for a rendering
of this paper.)

By the end of the 1950's the atmosphere in Cambridge had
changed. Physics had become a department and Dirac was thought
to be something of a relic. He was doing significant work but not in
the mainstream. He was ejected from his office and asked to teach
more. His parking spot was even taken away. Manci decided that
they had to leave. But Dirac stayed on until his official retirement in
1969. They spent the next two years going back and forth between
Britain and the United States. But then he got an offer to become
"Visiting Eminent Professor" at Florida State University in Talla-
hassee, which he accepted. He did some semi-popular writing and
gave lectures at various places. Dirac was a good lecturer. He made
meticulous preparations and did not deviate from his text. Questions
were not encouraged. There is a famous story about a Dirac lecture.
There was a question period and one of the attendees said that he

did not understand a certain equation in the lecture. Dirac made no response. After a long pause, the person running the colloquium reminded Dirac that he had been asked a question. Dirac replied, "It wasn't a question."

These various stories exemplify the question that we started out with. Where in the spectrum of eccentrics should we situate Dirac? Farmelo has an answer. He thinks that Dirac was autistic — all the symptoms he feels are there. Maybe he's right and maybe he isn't. In a way it hardly matters. A few years ago a colleague of mine named Pierre Ramond told me a story about Dirac which I found almost unbearably sad. Ramond is a professor of theoretical physics at the University of Florida in Gainesville. He gave a lecture at Florida State. Afterwards he invited Dirac to come to Gainesville to give a talk, even offering to drive him both directions. Dirac answered, "I have nothing to talk about. My life has been a failure." Ramond thought that if Dirac's life had been a failure, what does this say about the rest of us?

After a prolonged period of illness Dirac died on the 20th of October 1984. He was buried in the Roselawn Cemetery Talahasse, Florida. In 1995, after a considerable campaign he was commemorated in Westminster Abbey. The dean had initially objected on the grounds that Dirac was an atheist. It took until 1995. One cannot imagine that if Dirac had had a vote while he was still alive, he would have cared.

Appendix: The Other Dirac Equation

The purpose of this appendix is to present a derivation of Dirac's equation for the maximum separative power that a given centrifuge can have. Dirac did this work early in the war. It became the basis of the theory of gas centrifuges, the kind that are used to separate isotopes such as the isotopes of uranium. This is what motivated it. To do useful separation of isotopes many centrifuges must be linked together in a cascade but this equation tells us (or can tell us) the minimum number of centrifuges that are required to perform a certain task such as enriching uranium isotopes — that is changing the concentration of the light isotope uranium-235 in a mixture that contains mostly uranium-238. This mixture is combined with fluorine to make uranium hexafluoride gas which is introduced into the centrifuge. Real centrifuges are less efficient than the idealized model centrifuges in this derivation, just as real engines are less efficient than the engines of the Carnot cycle. Dirac's equation and the Carnot cycle are useful in the same ways. Dirac's five page unpublished paper is pretty laconic. The more recent treatments do things quite differently. They begin with Dirac's "value function" that defines the separative work that a centrifuge does and work backwards to find the maximum. But it was in this paper that Dirac first introduced the separative work function, after he had presented the argument that I am going to explain.

In a gas centrifuge two forces are in play. There is the centrifugal force which pushes the molecules towards the wall of the rotor and there is the hydrodynamic force that opposes it. Each of these forces produces a current and when these currents balance at some location in the gas we say that we have "equilibrium" at that location. In equilibrium no separation takes place. But nonetheless it is useful to analyze this case. With the assumption of equilibrium we can find the partial pressures p_1 and p_2 of two gasses, one which is composed of a light isotope and one which is composed of a heavier isotope. We can then find the ratio of these pressures which will give us the ratio of the mole fractions. This ratio is called the "separation factor" and is usually called α. It is close to unity in practical examples. As we shall see, this quantity enters the expression for the separative power. Call r the radial distance measured from the center of the rotating cylinder. In equilibrium due to the pressure is a function only of this r. The equilibrium condition is that the hydrostatic force due to the pressure difference at two nearby values of r is equal to the centrifugal force. We shall call the height of the gas h. Thus the hydrostatic force is given by

(1.) $$2\pi rh(p(r + \delta r) - p(r)) \approx 2\pi rh\delta r dp(r)/dr.$$

But in equilibrium this must equal the centrifugal force $2\pi rh\delta r\rho\omega^2 r$ where ω is the rotational frequency. For an ideal gas we have

$\rho = (M/RT)p$. Thus we have the equation

(2.) $$(1/p(r))dp(r)/dr = (M/RT)\omega^2 r.$$

Using the ideal gas law we can also write this as an equation for the density. We would therefore conclude that the change in the density is proportional to the density. In his brief paper, "The Theory of the Separation of Isotopes by Statistical Methods,"[j] Dirac informs us that this result could have been anticipated. In his paper he considers a quantity he calls "c" which is the ratio of the number of atoms of the lighter isotope to the total number of atoms. He writes,

"If the fluid at any point is in equilibrium under the action of the centrifugal force or temperature gradient, there will be a certain gradient in c, at that point. This gradient will be in the direction of the centrifugal force or temperature gradient. Its magnitude will be proportional to c, since the various light molecules (molecules containing the light isotope) move independently of one another, owing to their concentration being small..."

Equation (2) can readily be integrated to find

(3.) $$p(r) = p(0)\exp(M\omega^2 r^2/2RT).$$

If we have a two-component gas, say the isotopes uranium-235 and uranium-238, then we can use this expression to get the ratio of the partial pressures; i.e.

$$p(r)_1/p(r)_2 = p(0)_1/p(0)_2 \exp((M_1 - M_2)\omega^2 r^2/2RT)N(r)_1/N(r)_2$$
(4.) $$= N(0)_1/N(0)_2 \exp((M_1 - M_2)\omega^2 r^2/2RT)$$

From this we see that, taking r to be at the perimeter of the centrifuge

(5.) $$\alpha = \exp((M_2 - M_1)v_p^2/2RT),$$

where v_p is the peripheral speed.

It is instructive to evaluate the argument of the exponential in a typical case. For modern centrifuges $v_p \sim 600$ meters/second.

---

[j]This paper, minus an appendix that Dirac refers to, is held in the Public Records Office at Kew, London. The status of the appendix is unknown.

The mass difference in moles is about 3. T is about 300 and R = 8.3×10^7 so

(6.) $$(M_2 - M_1)v_p^2/2RT) \sim 0.2$$

As I have mentioned, Dirac considers the quantity $c = N_1/(N_1 + N_2)$. But he also makes the assumption that $N_1 \ll N_2$ such that $c \sim N_1/N_2$ so our previous work is relevant.

In general the gradient of the concentration c will not have its equilibrium value. But it will be proportional to c. Let us call the vector proportionality factor **B**. Suppose we have a small slab of width b with its face oriented normal to this vector. Then the change in c across the slab is bBc where B is the length of the vector **B**. Net diffusion of the light isotope only happens because the gradient of c is not the equilibrium gradient **A**c. So the net diffusion will be proportional to $c(\mathbf{A}-\mathbf{B})$. Thus the diffusion of the light isotope across the slab is given by $|\mathbf{B}|cbx\rho D|\mathbf{A}-\mathbf{B}|cS\cos(\theta)$. Here ρ stands for the density, S is the area of the side of the slab normal to **B**, and θ is the angle between **B** and the direction of the diffusion.

Dirac is looking to maximize this expression. That will occur when $\cos(\theta) = 1$, when the diffusion vector is normal to the surface of the slab. This leaves us the job of maximizing $B(A-B)$ which occurs for $B = 1/2\,A$. Thus we can make use of our work for the equilibrium case. The magnitude of the equilibrium diffusion current is, with D the diffusion coefficient measured in $(\text{length})^2/\text{sec}$ and ρ the mass density,

(7.) $$J = -D\rho dc/dr \sim D\rho c(M_2 - M_1)\omega^2 r/RT$$

The maximum amount of the light isotope transferred per unit area of the slab face by diffusion is the current multiplied by the change in concentration which gives $D\rho b((M_2 - M_1)\omega^2 r/2RT)^2 c^2$, taking into account the factor of $\frac{1}{2}$ discussed above. If we multiply this by the area of the slab face S then we have the total deposit. The separation power per unit volume is defined by Dirac to be the coefficient of c^2 in the above expression; i.e., $D\rho((M_2-M_1)\omega^2 r/2RT)^2$ bS is the maximum separation power of the volume bS. If we want the maximum separation power, U, of the entire cylindrical centrifuge of radius r_p we have

$$U = D\rho((M_2 - M_1)/2RT)^2 \omega^4 2\pi h \int_0^{r_r} r3dr$$

(8.) $= \pi/2D\rho((M_2 - M_1)/2RT)^2 hv_p^4.$

This equation, with $\Delta M = M_2 - M_1$, reads

(9.) $U_{max} = \pi/2D\rho(\Delta M/2RT)hv_p^4$

which says that the separative power of an ideal gas centrifuge goes as the length of the centrifuge and the fourth power of the peripheral velocity. This is Dirac's other equation. The efficiency of a centrifuge is defined to be the ratio of its actual separation power to the ideal. Real centrifuges can run with an 80 to 90 percent efficiency and consume about forty watts of electric power — roughly equivalent to a dim light bulb.

In actual centrifuge operations a fraction θ is produced as enriched product and a fraction $1 - \theta$ is recycled as unenriched waste. The separative power U is given in terms of θ and α as

(10.) $U = L[\theta/(1 - \theta)]\frac{1}{2}(1 - \alpha)^2 \sim L/2(1 - \alpha)^2,$

when $\alpha \sim 1$ and $\theta \sim 1/2$. L is a constant that characterizes the centrifuge. Finally, this whole discussion is premised on the proposition that the density of the gas is a continuous function of ones, position in the centrifuge. But in actual operation there is a critical speed at which the gas becomes pushed away from the interior and clusters on the inner surface. In this case it can be shown that the separative power has a dependence v_p^2 rather than v_p^4. I do not know which case applies to which actual centrifuges in current use.

5. Another Dirac

"A description of the proposed formulation of quantum mechanics might best begin by recalling some remarks made by Dirac concerning the analogue of the Lagrangian and the action in quantum mechanics. These remarks bear so directly on what is to follow and are so necessary for an understanding of it, that it is thought best to quote them in full even though it results in a rather long quotation."

Richard Feynman[a]

"This note is concerned with methods of separating isotopes which depend on subjecting the mixture of isotopes, in liquid or gaseous form, to physical conditions which tend to cause a gradient in the concentration of an isotope. The most useful examples of such physical conditions are the presence of a field of centrifugal force or of a temperature gradient. There is a general theory governing the performance of a separator which employs such a method. It puts an upper limit to the output of the apparatus and shows what running conditions one should strive to attain in order to approach the theoretical limit in practice."

P.A.M. Dirac[b]

---

[a] *Feynman's Thesis: A New Approach to Quantum Theory*, edited by Laurie M. Brown, World Scientific, New Jersey, 2005, p. 26.

[b] This paper, "The Theory of the Separation of Isotopes by Statistical Methods," was never published by Dirac but is held in the Public Records Office in Kew, London. I am grateful to Helmut Rechenberg for supplying a copy.

1. Paths

I would imagine that if the average physicist was asked to list Dirac's achievements in physics he or she would say "the Dirac equation." A few might say "the Dirac delta function" but this is a mathematical convenience and not exactly a discovery in physics.[c] Some might also say Fermi–Dirac statistics but then wonder why the alphabetical order of the names had been reversed. Dirac himself made this clear when he wrote in his quantum mechanics text, "This [the Pauli principle] is an important characteristic of particles for which only antisymmetrical states occur in nature. It leads to a special statistics which was first studied by Fermi, so we shall call particles for which only antisymmetrical states occur in nature *fermions*."[d] Of course Dirac made important contributions to the formulation of quantum theory, some of which I will shortly discuss, but in awarding him the Nobel Prize in 1933 — which he shared with Schrödinger — these were not discussed least of all by Dirac. His Nobel lecture was entirely concerned with the Dirac equation and anti-matter. As I said previously, I believe that if Carl Anderson had not discovered the positron in 1932 Dirac would not have gotten the prize at that time. In this note I am going to discuss two of Dirac's contributions that for one reason or another are not discussed as frequently. The first will be for an obscure paper in which Dirac presented the ideas that led to the path integral formulation of quantum mechanics which was first exploited in Richard Feynman's PhD thesis. The second will be Dirac's theory of the separation of isotopes using the gas centrifuge which I also discussed in the previous essay. This work was done at the beginning of the Second World War and introduced ideas that have been the basis of the subject eversince.

It appears that of Dirac's quantum mechanics papers, his 1927 paper was his favorite. Its elegant mathematics appealed to him greatly. Not only did he introduce the delta function but also the terminology "c-number" and "q-number" to distinguish between

---

[c]Dirac introduced it in a paper entitled *The Physical Interpretation of the Quantum Dynamics*, Proc. R. Soc. Lond. A. 1927, **113**, 621–641.

[d]*Quantum Mechanics*, by P.A.M. Dirac, Oxford University Press, Oxford, 1947, p. 210.

classical ordinary numbers and quantum mechanical operators. He introduced the notation (ξ'/η') to represent matrix elements. The purpose of this paper was to show how the notion of the canonical transformations of classical Hamiltonian dynamics is realized in the quantum theory. It will be recalled that in classical mechanics it is sometimes useful to replace the coordinates $q(t)$ and the momenta $p(t)$ by new coordinates $Q(q, p, t)$ and $P(q, p, t)$ such that Hamilton's equations, where H is the Hamiltonian,

(1.) $$dp/dt = -\partial H/\partial q$$

and

(2.) $$dq/dt = \partial H/\partial p$$

are preserved. This imposes special conditions on the transformation which need not concern us here. The quantum mechanical version of these canonical transformations is a transformation of the operators. Using Dirac's notation, if g is the operator being transformed, G the result and b the operator generating the transformation then

(3.) $$G = bgb^{-1}.$$

This transformation preserves the canonical commutation relations. If g is Hermitian and we want to preserve this feature then we require that $b^{-1} = b^{\dagger}$ where $b^{\dagger}$ is the hermitian conjugate of b. Dirac did not in this paper use the term "unitary" for this type of transformation. Dirac wanted to explore the unitary transformations that diagonalize g. If g is the Hamiltonian then these diagonal elements are the allowed energies of the system.

He considered two operators with matrix elements

(4.) $$\xi(\xi'\xi'') = \xi'\delta(\xi' - \xi''),$$

and

(5.) $$\eta(\xi'\varsigma'') = -ih\frac{\partial}{\partial\varsigma'}\delta(\xi' - \xi'').$$

One readily shows that these operators are canonically conjugate. Dirac then considered any function of these operators, $F(\xi, \eta)$. One wants to find a canonical transformation that diagonalizes this function. That is, we want to reduce it to the form

(6.) $F(\alpha'\alpha'') = \delta(\alpha' - \alpha'')F(\alpha')$.

In other words we need the matrix elements (ξ'/α') that accomplish the transformation. Dirac shows that these matrix elements obey the ordinary differential equation,

(7.) $F(\xi', -ih\partial/\partial\xi')(\xi'/\alpha') = F(\alpha')(\xi'/\alpha')$.

The $F(\alpha')$ are the diagonal matrix elements. He then noted that if the variables are identified with the coordinates and momenta q and p, and if F is the Hamiltonian, then the above equation is the time independent Schrödinger equation. Here the $F(\alpha')$ are the energies and the (ξ'/α') are the Schrödinger wave functions which in this view of things diagonalize the Hamiltonian. This rather straightforward argument demonstrates that the Schrödinger and Heisenberg pictures are simply two different representations of the same theory. This had already been claimed in a long and rather obscure paper by Schrödinger.[e] Curiously Dirac makes no reference to this paper although he does refer to a Schrödinger paper that had been published later. One wonders if he read the earlier paper or if he decided that it was irrelevant. His own argument is a masterpiece of economy. Jordan also discussed the transformation theory but his notations are pretty opaque compared to Dirac's.

I think that a fair summary of Dirac's work on the quantum theory to this point is that while it is very impressive, with the exception of the introduction of the Poisson brackets, it was work that in one form or another had also been done by others. The "Dirac equation" which Dirac formulated in 1928 is something else. It is a work of inspired originality and it is for this that Dirac won the Nobel Prize. It is the same kind of originality that characterizes Dirac's work on the Lagrangian in quantum mechanics although it took some time for it to be appreciated. The reason for this was partly the odd way it was published which goes back to Dirac's nature. He was not a person who needed a great many human contacts. But like many solitary people those he had ran deep. One of them was with the

---

[e] *Annalen der Physik*, 4, **79**, 1926, 734–756.

aforementioned Russian physicist Kapitza. Prior to the publishing of his paper in 1933 Dirac had made some visits to the Soviet Union and had even done some mountain climbing there. The only sport that Dirac had any interest in was rock climbing. Thus Dirac chose to publish his paper, "The Lagrangian in Quantum Mechanics" in the now long-defunct Russian journal, *Physikalische Zeitschrift der Sowjetunion*.[f] This practically guaranteed that the paper would not be widely read. Dirac also published the basic ideas in his quantum mechanics text. However they are buried in the middle of the book and easily skipped over. In 1941 Feynman was looking for a way to quantize theories where there was no classical Hamiltonian. At this time Herbert Jehle was a visitor to Princeton and he called Feynman's attention to Dirac's work. Since Feynman never cited Dirac's paper but only the book my guess is that he never read the paper.

Dirac began his paper by explaining why it was natural to quantize classical theories using Hamiltonian dynamics. Once you know how to represent position and momentum as operators it is elementary to represent any function of them such as the Hamiltonian as an operator. But this limits one's options. The Hamiltonian is not a relativistic invariant so this formulation is intrinsically non-relativistic. On the other hand the action S is a relativistic invariant so if it could be used one broadens one's possibilities. The key idea is the observation that going from a basis in which the coordinate $q(t)$ is diagonal to one in which q at a different time, say T, is diagonal can be achieved by a canonical transformation in which the generating function is the action. Dirac made the absolutely remarkable statement:

$$\text{``}\langle q_t \mid q_T \rangle \text{ corresponds to } \exp\left[i \int_T^t L dt/h \right].\text{''}$$

What can "corresponds to" possibly mean? He did not explain either in his paper or his book. Feynman took this to mean that one must be proportional to the other and worked out the proportionality factors in some examples. He once asked Dirac if he had ever done the same and was told "no."

---

[f]Band **3**, Heft 1 (1933), pp. 64–72. It is reproduced in Brown op. cit.

Dirac now considered the case where T and t differ from each other only infinitesimally and he made a similar statement,

"$\langle q_{t+dt} \mid q_t \rangle$ corresponds to $\exp[iLdt/h]$."

Again there is no explanation of what "corresponds to" means. In ordinary quantum mechanics if one writes the solution to the time dependant Schrödinger equation as $A \exp(i/\hbar S)$ then under reasonable assumptions S obeys the Hamilton–Jacobi equation $\partial S/\partial t = -H(q, \partial S/\partial q)$ where H is the Hamiltonian. Thus S is the action. Dirac's relationship is a generalization of this result as he noted in his paper.

Dirac now imagined dividing the time interval between T and t into many short intervals. He then choose "paths" between the q's at these times. He writes,

$$(8.) \quad \langle q_t | q_T \rangle = \int \langle q_t | q_m \rangle dq_m \langle q_m | q_{m-1} \rangle dq_{m-1} \cdots \langle q_2 | q_1 \rangle dq_1 \langle q_1 | q_T \rangle$$

where the integral is a multiple integral over all the intermediate q's; i.e. a "path integral." Each of these scalar products will take the form that Dirac proposed in terms of an exponential of the action. Then Dirac asked what is the classical limit of this expression? Each of the scalar products involves the integral of the Lagrangian divided by $\hbar$. In the classical limit $\hbar$ goes to zero. The integrands wildly oscillate with the exception of those paths in which the action is stationary. But these are the classical paths and these are the ones that contribute to the expression in that limit. He closed his paper with some general remarks. One looks in vain for any application.

Feynman never published his thesis. Soon after he wrote it he and his advisor John Wheeler went off to war. After the war Feynman published some of the results in an article in *The Reviews of Modern Physics*,[g] "Space-Time Approach to Non-Relativistic Quantum Mechanics." In an appendix I will present a calculation of this probability amplitude in the simplest case where there is no interaction. Other cases like the harmonic oscillator can be done but more generally it is a difficult method to apply. Nonetheless the

---

[g] *Rev. Mod. Phys.* **20**, 367–387 (1948). Both this paper and the thesis are included in Brown op. cit.

formalism and its descendants are at the heart of modern discussions of the quantum theory whether in the guise of "paths" or "worlds" or "histories." It is interesting to reflect that all of this can be traced back to an obscure paper by Dirac.

2. Separations

Dirac's interest in the separation of isotopes went back to the early 1930's. It was an active subject in Cambridge where Francis Aston, the inventor of the mass spectrograph, for which he won the 1922 Nobel Prize in chemistry, was a professor. Dirac proposed a method, the generic name for which is a "stationary centrifuge." Here the gas to be separated into its isotopes moves while the object that does the separation remains stationary. Dirac's idea was to force the gas to move through a large angle in a bent tube. The heavy component would be bent less. He actually carried out an experiment much to the amusement of Rutherford. The results were hard to interpret. He was going to carry out more with Kapitza, but Kapitza was detained in Russia and Dirac dropped the matter. It was taken up again during the Second World War and the South Africans used a version of the stationary centrifuge to separate enough uranium isotopes to make several nuclear weapons which they destroyed without testing in the 1990's.

In March 1940 Rudolf Peierls and Otto Frisch produced the memoranda that started the British nuclear weapons program and to a certain extent the Americans. (See the next essay.) It was immediately clear that the *sine qua non* was the separation of uranium isotopes. Dirac was contacted and he began his wartime activity devoted to isotope separation. It seems that it was in 1941 that Dirac wrote his seminal paper "The Theory of the Separation of Isotopes by Statistical Methods." This paper got to Peierls who was working at that time with Klaus Fuchs who even then was spying for the Russians. Peierls and Fuchs produced the standard paper on isotope separation[h] which found its way into the American work at the

---

[h] "Separation of Isotopes" by K. Fuchs and R. Peierls, in *Selected Papers of Sir Rudolf Peierls* edited by R. H. Dalitz and Sir Rudolf Peiels, World Scientific, Singapore, 1997, 303–320.

hands of people like Karl Cohen. These people all credit Dirac with the basic ideas.

Before I describe some of Dirac's contributions to the theory of isotope separation by centrifuges — especially gas centrifuges — I need to describe briefly how such centrifuges work. They consist of cylinders some ten to twelve centimeters in diameter and a meter or two in length. Because the details of modern centrifuges are classified one cannot get precise specifications. The best modern centrifuges are made of carbon fiber and can rotate around their long axes at peripheral speeds of some 700 meters per second. Before the centrifuges begin to rotate they are put into a vacuum to eliminate air resistance. Once they are rotating the gas is introduced. For the separation of the uranium isotopes U-235 and U-238 the gas that is used is uranium hexafluoride. The gas acquires the rotating motion of the cylinders. The heavier isotope is pushed more readily to the centrifuge wall by the centrifugal force. This is how the separation is produced. An obvious question to ask in this case is, won't the isotopes become completely separated if you leave the gas in long enough? To see why this is not the case an analogy is useful: the gravitational separation of isotopes in the stratosphere.

Consider a rectangular slab of atmosphere. If its total mass is m then a gravitational force of mg is pulling it down towards the earth's surface. But due to the difference in pressures at the top and bottom of the slab assuming that the density falls off as the distance above the earth's surface increases, there is a net upward pressure which balances the force of gravity. The difference in pressure is at equilibrium equal to the downward gravitational force, i.e., $dp = -g\rho dz$ where ρ is the density of the gas and z is the height above the earth. If we assume that the gas is ideal and that the temperature remains the same throughout the slab — something that is not actually true for the atmosphere — we have the equation, with μ the molar mass and R the gas constant,

(9.) $$dp = -g\mu p/RT dz,$$

or

(10.) $$dp/p = -g\mu/RT dz$$

which integrates to

$$(11.) \qquad p/p_o = \exp -((g\mu/RT)z)$$

which is the "barometric formula".

In 1919 Frederick Lindemann and Francis Ashton published their seminal paper entitled "The Possibility of Separating Isotopes."[i] Ashton, as I have mentioned, won the 1922 Nobel Prize in Chemistry for his use of the mass spectrograph to separate isotopes and Lindemann became Churchill's war-time science advisor and later Lord Cherwell. In this paper they explore various separation methods, one of which is the use of gravitation. They say that starting at a certain height above the earth, h_o, the isotopes of say neon, which is the case they study, will no longer mix by convection so that can be separated gravitationally. If you call the density of the heavy isotope ρ_1 and the density of the light one ρ_2, then with the assumptions that led to the barometric formula

$$(12.) \qquad \rho_1/\rho_2 = (\rho_1/\rho_2)_0 \exp(-(g\Delta h/RT(\mu_1 - \mu_2))).$$

Here Δh is the height above which the convection mixing stops. Aston and Lindemann suggest designing a balloon that would rise to 100,000 feet where it could sample the ambient atmosphere and look for isotopes of neon. But they conclude, "Although the quantities are measurable they do not appear sufficiently striking to warrant the outlay and labour such experiments would entail."[j] There appear to be some experiments on South Polar ice that show evidence for this kind of gravitational separation. More relevant to us is what Aston and Lindemann have to say about centrifuges.

They argue that the same equation holds if you substitute the centrifugal acceleration $v^2/r = \omega^2 r$ for the gravitational acceleration g. At the edge of the centrifuge the ratio of densities would be

$$(13.) \qquad K/K_o = \exp(-v^2/RT(\mu_1 - \mu_2)).$$

---

[i] *Phil. Mag.*, Vol. xxxvii, p. 523, 1919.

[j] Lindemann and Aston, "The Possibility..." op. cit. p. 531.

Here v is the peripheral velocity and K is the density ratio at the edge while K_0 is the density ratio at the center. They put in some numbers and concluded that a peripheral velocity of at least a thousand meters a second would be required to make useful separations. In 1934 J. W. Beams and F. R. Haynes used a centrifuge to separate the isotopes of gaseous chlorine. It had a maximum peripheral velocity of some 800 meters a second before it burst. Commercial gas centrifuges with modern materials can run at these speeds.

Given the assumptions, this formula is a useful way to estimate the percentage separation that a given centrifuge can perform. What it does not tell us is the rate at which this can be done. A theoretical maximum was supplied by Dirac in his paper. One should think of this the way one thinks of the Carnot cycle. The Carnot cycle provides the optimal performance of an ideal heat engine. Real heat engines will perform less well compared with real centrifuges. In a previous essay I have presented a derivation of this maximum.[k] The result is for a cylindrical centrifuge of length h is in kilograms per second

$$(14.) \qquad U_{max} = \pi/2 D\rho(\Delta\mu/2RT)^2 hv^4.$$

Here ρ is the density of say the light component of the gas, v is the peripheral velocity and D is the diffusion coefficient measured in meters squared per second. The dependence on the fourth power is very striking but for actual centrifuges it is more like the second power. My concern here however is Dirac's introduction of the unit he called "sep-power." This is a measure of how much separation power is needed to perform some given task such as producing a kilogram of 90% enriched uranium starting with uranium hexafluoride which used natural uranium with a concentration of about 0.7% uranium-235. The Dirac "value function" is used to calibrate this effort.

I am not going to present Dirac's derivation — at least as it is presented in the available part of his paper — since it assumes that the isotopic concentrations are all small. The paper refers to an appendix which has never been made available where this restriction

---

[k]*American Journal of Physics*, **77**, (2009) 979–987.

is dropped. Rather I will present the derivation given by Peierls and Fuchs in 1942. They refer to Dirac's paper and presumably they saw the appendix.

When there is isotope separation there is an entropy change ΔS. Peierls and Fuchs label the concentration of the light isotope c and therefore the concentration of the heavy isotope is $1 - c$. They define a quantity F as

(15.) $$F = \Delta S/c(1 - c).$$

In a footnote they explain the denominator. "The reason for this is that, with all usual methods, the work done by the device on the molecules is approximately independent of their nature. Of all possible pairs of molecules only the fraction $c(1 - c)$ are unlike ones, and only on those cases can the work be done for the purpose of distinguishing them lead to any discernible result. In all other cases it is wasted. Hence the factor $c(1 - c)$ in the efficiency."[1] They define a quantity

(16.) $$\Delta Y/\Delta t = R\Delta F/\Delta t$$

as the separating power. To find this we need an expression for the change in entropy ΔS produced by a separation of two constituents in a binary mixture. ΔS is given by

(17.) $$\Delta S = -R(c\ln(c) + (1 - c)\ln(1 - c)).$$

If you introduced a semi-permeable membrane into the original volume and, if there was a fifty-fifty admixture of the two components ($c = 1/2$), then after complete separation, the total entropy of the separated components would be

(18.) $$S = k\ln(2^N)$$

where N is the total number of molecules. If you assume a small change in the concentration, 'd' and expand in a Taylor series, this

---

[1] Fuchs, "Separation of Isotopes" op. cit. p. 303.

would produce a small change in the entropy, δS, given by

(19.) $$\delta S \sim Rd^2/2c(1-c),$$

or

(20.) $$\Delta Y = Rd^2/2(c(1-c))^2.$$

They introduce a quantity $y(c)$ which represents a measure of the total effort to produce one mole of concentration c from an ordinary mixture of isotopes. The dimensionless quantity they define as the "separation potential." ΔY can be expanded in a Taylor series and because of conservation laws the first non-zero term is the coefficient of the second derivative of y with respect to c. Hence cancelling terms one is led to the differential equation

(21.) $$d^2/dc^2 y(c) = 1/(2c^2(1-c)^2),$$

which has the solution

(22.) $$y(c) = (2c-1)\log(c/(1-c)) + ac + b,$$

where a and b the integration constants. Following Dirac, they fix these constants by insisting that if c_0 is the concentration of one of the components of the natural mixture then both y and its derivative must vanish at $c = c_0$. This gives them a form of the function

$$V = (2c-1)\ln((c/1-c))(1-c_c)/c_0))$$
(23.)
$$+ (c - c_0)(1-2c_0)/c_0(1-c_0).$$

However the common treatment sets $a = b = 0$. This leads to what is called the "Dirac value function" $V(c)$ where

(24.) $$V(c) = (2c-1)\log(c/(1-c)).$$

To see how this is used I am going to consider the case of the separation of uranium isotopes by centrifuge. A gas centrifuge has a portal for the feed and two portals for the output. Through one of these output portals passes the "product" — the unenriched uranium. Through the other passes the unenriched uranium or the "tails." In 1939 Harold Urey invented the idea of "counter currents." The heavier gas is made to move downward at the periphery while the light gas moves upward at the center. As one of his contributions Dirac worked out the basic theory of this which was the foundation of the future design. The operator of the centrifuge sets the percentage of the isotopes in the tails as well as that in the product. Given these percentages and the percentage in the feed one can use the Dirac value functions to evaluate the work needed to produce say a kilogram of uranium-235. The separative work unit is defined by the equation

(25.) $$\mathrm{SWU} = WV(x_w) + PV(x_P) - FV(x_F).$$

Here the various x_i's are the concentrations and the V's are the Dirac value functions. W, P and F are the quantities of waste, product and feed usually measured in kilograms. However what one does is to divide by P and write

(26.) $$\mathrm{SWU/kilogram} = W/PV(x_w) + V(x_P) - F/PV(x_F).$$

In this process the quantity of uranium is preserved which means that these ratios are fixed by the concentrations. If you set the product to be say one kilogram you have

(27.) $$\mathrm{W/kg} = (x_f - x_p)/(x_w - x_f),$$

(28.) $$\mathrm{F/kg} = (x_w - x_p)/(x_w - x_f).$$

This means that the SWU — "Swoo" — per kilogram can be readily computed by using one of the many SWU calculators that you find on the web. Here are a few samples.[m]

x_f	x_w	x_p	SWU/kg U^{235}
0.00711	0.0025	0.95	232.39
0.044	0.0025	0.95	72.46
0.199	0.0025	0.95	22.51

---

[m]These numbers are taken from a SWU calculator of R. L. Garwin. I thank him for making it available.

What is striking is how rapidly the SWU fall off as the feed sample becomes more enriched. The first case is natural uranium. The second is reactor grade uranium and the third is the upper limit of what is called low enriched uranium. We can understand the trend if we imagine looking for needles in a haystack — the needles being U-235. The more highly enriched the feed the more "visible" the needles and the easier our task is. To put these numbers in perspective, the Dirac limit for the kind of centrifuge that the Iranians have been employing is about five SWU per year although the actual SWU production is certainly substantially less. The best modern centrifuges can produce over a hundred SWU per year. An implosion weapon needs about 20 kilograms. It is also instructive to, say, double the waste concentration to 0.005. The SWU requirement drops to 172.41 to produce highly enriched uranium from natural uranium. A simple illustration might help to illuminate the issues. Suppose we want to produce a certain amount of orange juice. If the price of oranges is not an issue we can leave more waste orange after each squeezing, using less energy per orange but more oranges. When there is plenty of uranium hexafluoride available it might pay to increase the waste concentration. Again it was the work of Dirac that led the way.

Appendix

This appendix presents a calculation of the quantity $\langle x_t | x_o \rangle$ using path integrals in the simplest case possible where x propagates in time as a free particle. The calculation which I take from unpublished notes of M. Gell-Mann and M. L. Goldberger[n] already contains many features of the method. Doing realistic cases becomes very complicated.

If the particle propagates as a free particle its Hamitonian is simply $p^2/2m$. But the action is written in terms of the coordinates so we solve the Heisenberg equation for x_t to find

(1.) $$x_t = x_o + p_o t/m.$$

---

[n]I thank Murph Goldberger for sharing these notes.

It will be useful to evaluate the commutator of x_o and x_t. Using the previous equation

(2.) $$[x_0, x_t] = i\hbar t/m.$$

If we substitute the expression $p_0 = m/t(x_t - x_0)$ directly into the Hamiltonian we get terms involving both the product $x_0 x_t$ and $x_t x_0$ which are not the same. In the action we want to replace the operators by their eigen-values so we "well order" the Hamiltonian so that the x_t terms are always to the left of the x_o terms. Using the commutator the well-ordered Hamiltonian can be written as

(3.) $$H = m/2t^2\{x_0^2 + x_t^2 - 2x_t x_0\} - i\hbar/2t.$$

Replacing these by their eigen-values in the equation for $S(x_t, x_o, t)$ we have the Hamilton–Jacobi equation

(4.) $$-\partial S/\partial t = (m/2t^2)(x_t - x_0)^2 - i\hbar/2t.$$

This equation has the solution

(5.) $$S(x', x) = (m/2t)(x' - x)^2 + i\ln((\alpha t)^{1/2}).$$

Here α is a constant that is determined by the condition that as t goes to zero $\exp(i/\hbar(S(x', x)) \to \delta(x' - x)$. There is no transition. We can now use the expression

(6.) $$\delta(x) = \lim_{\varepsilon \to 0}\{(1/\sqrt{2\pi i\varepsilon}) \exp(ix^2/2\varepsilon)\}$$

to evaluate α. This gives us

(7.) $$(x'|x) = \exp(im(x' - x)^2/2\hbar t)/(2\pi i\hbar t/m)^{1/2}.$$

This object is referred to as a "propagator" since it propagates the state forward in time.

(8.) $$|x'\rangle = \int dx\langle x'|x\rangle|x\rangle.$$

6. More About Bohm's Quantum

"But in 1952 I saw the impossible done. It was in papers by David Bohm. Bohm showed explicitly how parameters could indeed be introduced into nonrelativistic wave mechanics, with the help of which the indeterministic description could be turned into a deterministic one, more importantly, in my opinion, the subjectivity of the orthodox version, the necessary reference to the 'observer', could be eliminated."

John Bell[a]

In the fall of 1951 I took my first real course in quantum mechanics from Julian Schwinger. I was a mathematics graduate student and a sort of disciple of George Mackey who had done work on the mathematical foundations of the theory. I knew a little about quantum mechanics from having taken a course in modern physics from Norman Ramsey and having done some study with Philipp Frank. I had also read some books on the subject but my overall knowledge was pretty skimpy. There was no textbook for Schwinger's course. He may have given a list of recommended reading but if so I have forgotten it. There were not that many texts available. There was Leonard Schiff's book *Quantum Mechanics*, first published in 1949, and David Bohm's book *Quantum Theory*, which was published in 1951. Both books

---

[a]J.S. Bell, *Speakable and Unspeakable in Quantum Mechanics*, Second Edition, (Cambridge University Press, Cambridge, 2004), p. 160.

interestingly had been inspired by the courses the authors had taken with Robert Oppenheimer. The use of "mechanics" in the one title and "theory" in the other is a good characterization. Schiff emphasizes how to do calculations and Bohm has a great deal of discussion about what the calculations mean. Abner Shimony, who was then a student of Wigner, asked Wigner what he thought of the book. He told Shimony he thought that it was a good book but that there was too much "schmoozing." He then asked Shimony if he knew what schmoozing was. In my case, and probably Shimony's, it was just the schmoozing that attracted us to the book. It appealed to Einstein too who informed Bohm that he thought that it was the best expression he had seen of the arguments against his critique of the theory. But then Bohm changed his mind.

In early 1952 Bohm's papers on his interpretation of the theory appeared. I think that the reactions of my mentors at Harvard was typical. They saw what Bohm had done as a very complex waste of time. I remember Wendell Furry noting with derision the immense complication of Bohm's discussion of the two-slit experiment as opposed to the simple quantum mechanical explanation. I too acquired this attitude not because I had made any real study of Bohm's papers, but because I did not see any motivation for studying them. This attitude persisted for many years until I began discussing quantum mechanics with John Bell. After a typical discussion I could sympathize with the Greeks who decided to induce Socrates to drink the hemlock. Bell had this extraordinary lucidity coupled with a vividness of expression. After an hour with him one wondered if one had any understanding of anything. But it was only in the years after his death that I returned to Bohm to see why Bell admired this work so much. Indeed Bell was convinced that the de Broglie–Bohm interpretation of quantum mechanics should be part of any college curriculum on the subject. Bell had never been part of a university department so he did not realize how hopeless this proposition was. Anyone who has taught quantum mechanics in such a department realizes that with the pressures of PhD qualifying exams and theses one barely has the time to squeeze in conventional quantum mechanics let alone some outré, unconventional interpretation. It is only many years later when doubts creep in about the foundations of the theory that one becomes receptive. At least this is what happened to me, although I was pushed in this direction by Bell.

Naturally I cast about for things to read on the subject. But nothing really satisfied me although Bell's essays came the closest. Many of the papers I tried to read were written by philosophers and had words like "ontology" sprinkled over them like paprika. Too much "schmoozing", as Wigner might have said. An exception is *The Undivided Universe* by Bohm and Hiley.[b] While there is a good deal of schmoozing in this book there is also a great deal of physics, albeit at a level that requires a good deal of background from the reader. So I decided to try to write something that I would have enjoyed reading as a beginner. I have to apologize in advance. I am not an expert. I have done no original research in this subject. I am an elderly student but having taught myself to some degree perhaps I can teach others.

Let me first begin by making clear what the de Broglie–Bohm theory is and what it isn't. The theory arose because de Broglie and then Bohm were unsatisfied by the lack of clarity of the role that the Schrödinger wave function plays in ordinary quantum mechanics although in de Broglie's case this is not exactly clear. If you read the transcript of his lecture on the subject in the 1927 Solvay Congress nothing is very clear. When the wave equation was first published Einstein was very enthusiastic. He together with Schrödinger for that matter, saw these waves as material oscillations in ordinary space-time. They might act like "Führungsfeld" — "guide fields" — Einstein's term — for the observed particles thus resolving the duality between the wave and particle nature of matter. It soon became apparent that this interpretation in terms of material oscillations in ordinary space did not work. Wave functions involving more than one particle were multi-dimensional so that their arguments were in "configuration space" and not real space. Max Born then gave the interpretation that we have had eversince: These are waves of probability amplitudes in configuration space. Nonetheless de Broglie soldiered on with the idea that these were guide waves in Einstein's sense. Among others, Pauli made very strong but actually

---

[b]D. Bohm and B.J. Hiley, *The Undivided Universe*, Routledge, New York (1993).

irrelevant objections and de Broglie abandoned his project and so it lay abandoned until Bohm revived it.

There are two things one must understand at the outset (others will emerge later). The first is that the theory is non-relativistic and the second is that it does not produce results that differ from those of conventional quantum theory. When unambiguous the predictions of the two theories are identical. The matter of relativity is deeper. As his ideas progressed Bell came to believe that the real problem with quantum theory is relativity. There is of course no problem in making the Schrödinger equation covariant. This leads as is well known to the Klein–Gordon equation or the Dirac equation or other equations for particles of higher spin. The problem is in the interpretation. How does one understand the instantaneous collapse of the wave function in a measurement, or the instantaneous spin correlations of two widely separated singlet state electrons in the double Stern–Gerlach experiment — something that Einstein characterized as a "spooky" action at a distance? While there may not be an outright contradiction with relativity there is something about this that makes one a bit queasy. That there is a tension between the two theories is perhaps not surprising. Relativity deals with objects that move very fast and the quantum theory deals with particles that are very small. The two theories don't quite fit together. There is a good deal of speculative work on this which I will avoid completely by dealing only with non-relativistic quantum mechanics. With this in mind I will now present Bohm's presentation using the notation and equations of his original paper.[c]

The Schrödinger wave function ψ can be written as

$$(1.) \qquad \Psi = \mathrm{R}\exp(i\mathrm{S}/\hbar).$$

Here R and S are real functions of space and time. Using the Schrödinger equation

$$(2.) \qquad i\hbar\partial\Psi/\partial t = -\hbar^2/2m\nabla^2\Psi + V(x)\Psi.$$

---

[c] *Physical Review*, **85**, 166–193 (1952).

we have

(3.) $$\partial R/\partial t = -1/2m[R\nabla^2 S + 2\nabla R \cdot \nabla S].$$

If we let the probability $\Psi^\dagger \Psi$ be called P then $R = P^{1/2}$, or from the above

(4.) $$\partial P/\partial t + \nabla \cdot (P\nabla S/m) = 0,$$

which suggests that if we identify $\nabla S/m$ with a velocity and call $j = P\nabla S/m$ the current, we have a continuity equation. We will soon see what velocity is to be so identified. There is a second equation that reads

(5.)
$$\partial S/\partial t + (\nabla S)^2/2m + V(x)$$
$$-\hbar^2/4m(\nabla^2 P/P - 1/2(\nabla P)^2/P^2) = 0$$

where we have made use of the Schrödinger equation with potential V.

So far there is no new interpretation of the quantum theory involved. All we have done is to replace the Schrödinger equation by two equations for the modulus and phase of the wave function. It is the next step that suggests this interpretation. To see this we define the "quantum mechanical potential" U(x),

(6.)
$$U(x) = -\hbar^2/4m(\nabla^2 P/P - 1/2(\nabla P)^2/P^2)$$
$$= -\hbar^2/2m(\nabla^2 R/R).$$

If we set h = 0, Eq. (5) is just the classical Hamilton–Jacobi equation which can be derived from Newton's law. We have here a sort of quantum mechanical Hamiton–Jacobi equation. But let us suppose that we take the continuity equation seriously and identify $\nabla S/m$ with $\dot{Q}$ the velocity of a real particle of mass m. Then the "quantum mechanical Newton's law" is

(7.) $$m\dot{Q} = -\nabla(V(Q) + U(Q)).$$

See the appendix for a derivation.

This is Newton's law with the extra quantum mechanical potential. This is the form of the equations of motion for the Bohmian particles that you will find in Bohm. Most other authors present a first order equation from which Eq. (7) can be deduced. This is what de Broglie did. Bell may have been the first of the modern authors to write the equations in this first order form. Thus in our previous notation the first order equation for the time evolution of Q is

(8.) $\dot{Q} = j(Q,t)/P(Q,t) = 1/m\partial/\partial Q(\text{Im}(\log \psi(Q,t)))$.

In writing the second part of this equation we have used the fact that j can be written as

$$j = 1/m\text{Im}(\Psi^\dagger(Q,t)\partial/\partial Q\Psi(Q,t)).$$

As a consequence of the continuity equation (4), the probability density of the particle position (or configuration) will be $|\Psi(t)|^2$ at all times t, assuming (as we do) that it is so distributed initially.

Before we launch into the deep ends of how Bohm's interpretation deals with the issues of quantum theory, let's begin with the simplest example imaginable. In this example I will set $\hbar = 1$. Our *modus operandi* is first to solve the Schrödinger equation for the ψ we are going to insert in Eq. (8). The solution to the Schrödinger equation in this case is a free particle solution in which the normalization is irrelevant since it cancels in Eq. (8). The solution to the Schrödinger equation in this case is $\psi(\text{x},\text{t}) = \exp(i(\text{px} - \omega\text{t})$. If we insert this into Eq. (8) we get $\dot{Q} = p/m$ which tells us that in this case the Bohmian particle moves in a straight line with no change of speed. A more interesting example to work out is the one dimensional simple harmonic oscillator. The wave function for the ground state is, with A the normalization

(9.) $\Psi = A\exp(-x^2/2a^2)\exp(-i\omega_0 t/2)$,

with a $= (\hbar/m\omega_0)^{1/2}$.

Here the factorization into the R and $\exp(iS)$ is given to us on a platter. But notice that in this case S is independent of x. This means that the particle momentum is zero! We are indeed dealing with a stationary state in the sense that there is no motion. To see that this is consistent with "Newton's law" the reader is invited to work out the quantum potential. You will find that this cancels against the classical potential leaving behind a constant term — the ground state energy $\hbar\omega/2$ — which does not generate a force since it has zero gradient. This picture may or may not agree with one's intuition, but intuition is hard to objectify. Before I take up the more general considerations beginning with what the theory has to say about quantum measurements and the uncertainty principle I would like to discuss another example after I give a bit of background.

In 1953 Max Born retired from the Tait Chair in Natural Philosophy which he had held for some seventeen years in Edinburgh. Some colleagues decided to make a little volume of essays dedicated to him — *Scientific Papers Presented to Max Born.*[d] There is a remarkable collection of authors, including people like Herman Weyl, Schrödinger, and Louis de Broglie. A strange addition is Pascual Jordan, who was an ardent Nazi. Considering that Born was forced out of Germany, Jordan's appearance here is a little odd. But I want to focus on two of the essays, one by Einstein and a reply by Bohm. I was a little reminded of the kind of debates that Einstein and Bohr had in the beginning when Einstein was claiming that quantum theory was actually wrong. Here he is claiming that Bohm's interpretation is actually wrong.

In Bohm's translation from Einstein's German, Einstein claims that there is a "well-founded requirement [in quantum mechanics] that in the case of a macro-system, the motion of the system should approach the motion following from classical mechanics."[e] Bohm argues that not only is this principle not true in general it is also not relevant for the specific case that Einstein considers — a particle in an infinite well. If we take the boundaries to be at 0 and L then the wave functions take the well-known form

$$\Psi(x,t) = (2/L)^{1/2} \sin(n\pi x/L) \exp(i\omega t).$$

The only thing we need to know about ω is that it is independent of x. In the conventional interpretation the energy levels are given by

$$E_n = n^2 \frac{h^2 \pi^2}{2mL^2}.$$

---

[d]I am very grateful to Arthur Fine for calling my attention to this book. The copy I have was published by Hafner, New York, in 1954.
[e]Born, op. cit. p. 14.

We can interpret this by saying that the momentum of a particle of mass m, p_n, is $\pm n\hbar\pi/L$. However the wave function above is not an eigenfunction of momentum. It is a linear combination of the two exponential plane waves that are eigenfunctions, one belonging to the positive momentum and one to the negative. The interpretation of this wave function that Einstein accepts is that it represents a statistical ensemble equally weighted with momenta in the positive and negative directions. If a measurement is made then half the time one will find the positive momentum and half the time the negative.

Born contrasts this with his reading of Bohm. As in the harmonic oscillator the factoring of the wave function is again given on a platter and again since S does not depend on the spatial variable the corresponding velocity vanishes. This Einstein objects to on the somewhat curious grounds that if you make the well macroscopic the quantum effects go away and the particle should have either the positive or negative momentum and not zero as Bohm would claim. This is certainly an odd version of the correspondence principle. Bohm argues that one cannot make a claim about the momenta in the absence of a measurement and indeed if one does a measurement one will find the expected quantum mechanical answer and not the zero momentum to which Einstein objected. This had better be the case since Bohmian mechanics and quantum mechanics should agree. It is also instructive to compute the quantum potential in this case. One finds just the energies given above which correspond to $p^2/2m$ where the p here is the momenta found in the measurement. There was no further colloquy between them but it is unlikely that any minds were changed.

An extremely interesting example is what is called the Aharonov–Bohm effect or for the more historically fastidious the Aharonov–Bohm–Ehrenberg–Siday effect.[f] Let us suppose we have

---

[f]See W. Ehrenberg, R.E. Siday. "The Refractive Index in Electron Optics and the Principles of Dynamics," *Proceedings of the Physical Society B*, **62**, (1949) 8–21. The Aharanov–Bohm reference is Y. Aharanov, D. Bohm, "Significance of Electromagnetic Potential in Quantum Theory," *Phys. Rev.*, **115**, (1959) 485–491. The Ehrenberg and Siday paper is concerned with the geometric optics of electrons. As an example they consider the set up we are about to describe with a magnetic field confined to a solenoid and a split electron beam passing around it. They show that the de Broglie waves are phase shifted by an amount that

a solenoid that produces a magnetic field confined to a horizontal cylinder. There is a source of electrons that sends a beam towards the cylinder. There is a beam splitter that produces two beams, one that goes around the cylinder one above and one below. The two beams recombine and the interference effects are detected by say a photographic plate. The essential thing to remember is that the beams never interact with the magnetic field B so the Lorentz force $F = ev \times B$ where e is the charge of the electron, v its velocity and B the magnetic field, never comes into play. This means that intuitively we might imagine that the presence of the magnetic field would not matter. In fact it does matter and that is what needs to be explained.

In general $B = \nabla \times A$, where A is the vector potential. In the region where $B = 0$, A does not in general vanish although B does. In classical physics we are accustomed to thinking that A cannot influence the physics if for no other reason than it depends on the choice of gauge. We will now see that in quantum mechanics things are quite different. An alternative to the method I am going to sketch can be found in the textbook *Modern Quantum Mechanics* by J. J. Sakurai.[g] It makes use of Feynman path integrals.

Since the Hamiltonian is $H = (p - A)^2/2m$, and A does not vanish outside the cylinder, the evolution of ψ is not the same in the absence of a magnetic field. In fact, as a consequence of the Schrödinger equation, the interference of the waves that have propagated above the cylinder and those that have propagated below is changed by a phase factor proportional to the magnetic flux in the solenoid. This is the Aharanov–Bohm effect.

The same phase shift appears in Bohmian mechanics. The electron — the Bohmian particle — obeys Newton's law. There are two forces involved, the Lorentz force and the force derived from the quantum potential. External to the solenoid there is no Lorentz force but there is a non-vanishing quantum potential which depends on A.

---

is proportional to the magnetic flux. They remark that it is curious that this phenomenon has to do with the flux and not a change of flux. The only use of quantum theory in the paper is to assign de Broglie wave lengths to the electrons.
[g]J.J. Sakurai, *Modern Quantum Mechanics*, Benjamin/Cummings, Reading, Massachusetts, 136–139.

The reader can consult the Bohm–Hiley book previously referred to to see a drawing of this potential for this case and the resulting particle trajectories. This explanation has the nice feature that the quantum mechanical nature of this effect is clearly isolated in the quantum potential.

Many physicists have come to believe that there is a "measurement problem", or more than one, in quantum mechanics. Let me begin by reminding you again of how measurements are described in the orthodox — some would say the "Copenhagen" — interpretation of quantum theory. In this view the world is divided into "systems" and "apparatus." In the simplest case the system — which is quantum mechanical — is characterized by a wave function. (In more complex cases it is characterized by a density matrix.) Ultimately, this interpretation insists that the apparatus which finally performs and records measurements on the system must be classical. There are pointers and spots on photographic plates. What is unclear — certainly Bohr never made this clear — is where the dividing line is. This lack of clarity infuriated Bell who as mentioned kept referring to Bohr as an "obscurantist." This is one aspect of the measurement problem. But as a practical matter experimenters are always on the classical side of the line so let us ignore this matter for the time being.

In addition to the wave function the system is, as we have previously noted, characterized by "observables" which are represented by Hermitian operators A, B, C etc. The state of the system ψ can be expanded in eigenfunctions of any of these observables; $\Psi = \sum_i c_i \Psi_i$, where the c_i are probability amplitudes that weigh the different eigenvalue possibilities. The apparatus acts like a projector. When the system is operated on by the apparatus one of these eigenfunctions in the sum is projected out and becomes the wave function that describes the system after the measurement. This projection which is irreversible takes place when the outcome is recorded by, say, a pointer reading or a spot on a photographic plate. It has been known since the work of von Neumann in the early 1930's that this projection cannot be described by a Schrödinger equation. There are various ways of saying this but perhaps the simplest is to say that the Hamiltonian in the Schrödinger equation is a linear operator which takes linear combinations of wave functions into linear combinations. Another way of saying it is that in the usual picture the time evolution of a wave function is given by the unitary operator $\exp(i\mathrm{Ht}/\hbar)$ and hence it cannot cause the kind of "collapse of the wave function" that is produced by a measurement. Von Neumann simply said that there were two kinds of time evolutions — one given by the Schrödinger equation and one not. Many people find this unsatisfactory which is what constitutes for them the measurement problem. I will describe the Bohmian approach to this matter.

We suppose that the observer and observed are characterized by configurations r and q respectively. The variable r might refer to the configuration of a pointer and q to the position of a particle. It is reasonable to assume that initially, the wave function is a simple product of the particle wave function φ with its coordinate q and the apparatus wave function η with its coordinate r. Thus

(10.) $$\Psi(q, r, 0) = \Phi(q)\eta(r).$$

As a model of measurement, let us suppose that whenever φ is an eigenfunction of the observable A with the eigenvalue α, $\Phi = \Phi_\alpha$, the after-measurement wave function $\Psi(q, r, t) = \Psi_\alpha$ is one that has the apparatus point to alpha. Then, for a general φ, $\Phi = \sum_\alpha c_\alpha \Phi_\alpha$, we obtain

(11.) $$\Psi(q, r, t) = \sum_\alpha c_\alpha \Psi_\alpha.$$

The fact that we obtain a superposition of measurement results is the measurement problem. Now in Bohm's theory the configuration is $|\Psi|^2$ distributed. Thus, with probability $|c_\alpha|^2$ the apparatus is in a configuration pointing to α. No wave function has collapsed.

In the second of his two 1952 papers,[h] Bohm discussed measurement in the context of his interpretation. Von Neumann is mentioned but only in connection with his argument that with his assumptions a theory like Bohm's was impossible. Bohm's discussion is a

---

[h]David Bohm, A Suggested Interpretation of the Quantum Theory in Terms of "Hidden Variables", *Phys. Rev.* **85**, 180–193 (1952).

little vague and it was Bell who first sorted this out. One may wonder where the quantum mechanical uncertainties are. To understand this keep in mind that the initial conditions determine the trajectory. But these are subject to the usual quantum mechanical uncertainties. The motions of the particles on their trajectories are deterministic but which trajectory the particle is on is subject to probabilities.

I want now to turn to one of the most significant and characteristic aspects of Bohmian mechanics — its non-locality. This manifests itself first when there are two particles in interaction. In this case the wave function takes the form $\psi(r_1, r_2, t)$. Note that there is only one time as the theory is non-relativistic. Now there is the Schrödinger equation,

$$i\hbar \partial \Psi(r_1, r_{2,t})/\partial t = -\hbar^2/2m \nabla_1^2 \Psi(r_1) - \hbar^2/2m \nabla_2^2 \Psi(r_1, r_2, t)$$
$$(12.)\qquad\qquad + V(r_1, r_2, t)\Psi(r_1, r_2, t).$$

This leads to two equations, one for particle 1 and the other for particle 2, i.e.,

$$(13a.)\qquad \dot{Q}_1 = 1/m \partial/\partial Q_1 (\mathrm{Im}(\log \psi(Q_1, Q_2, t))),$$

and

$$(13b.)\qquad \dot{Q}_2 = 1/m \partial/\partial Q_2 (\mathrm{Im}(\log \psi(Q_1, Q_2, t))).$$

The thing to notice here is that solving for one particle also uses the instantaneous position of the other. This is the non-locality. Bell showed that non-locality is an ineluctable aspect of quantum theory. Here it is in your face while in the standard theory it is concealed. This kind of non-locality is just what Einstein objected to. He did not like Bohm's model but as we have seen he had other reasons.

Many of the ideas just articulated come into play in the Bohmian discussion of spin. The Bohmian particles do not carry spin so how can we describe, say an electron? To understand this we must think back to how we know that the electrons have spin. One bit of evidence is the Stern–Gerlach experiment. But this experiment does not measure "spin." It measures spots on a photographic plate. These show up in two separated lines which we explain by saying that the spin-1/2 particle has interacted with the Stern–Gerlach magnetic field which has split the incident beam in two. Therefore what we must show is that this is reproduced in Bohmian mechanics. I will just give the outlines of the argument. The interested reader can find the details in Bell.[i] We first solve the Schrödinger equation for a charged particle with spin interacting with a magnetic field. We then use this solution to construct the Bohmian current. Then we put this current into the equation for the trajectories as in (8). For the Stern–Gerlach experiment we find that there are two groups of trajectories, each group with a probability given by the Born rule. By looking at the spots on the photographic plate you cannot tell which interpretation of the quantum theory is in play. Indeed you can never tell. That is how the theory was designed.

It is nonetheless instructive to outline the usual method of describing this experiment. It will show us how the measurement problem manifests itself in this situation. I will follow the discussion in Bohm's text.[j] The particle with spin $\frac{1}{2}$ is subject to an interaction Hamiltonian given by

$$(14.) \qquad H_I = \mu(\sigma \cdot gH).$$

Here $\mu = -eh/2mc$, σ is the Pauli spin matrix and H the magnetic field. In the Stern–Gerlach set up H is inhomogeneous and points in the z direction. Expanding

$$(15.) \qquad H_I \approx \mu(H_0 + zH_0')\mu.$$

Before the particle passes the magnetic field the wave function takes the form

$$(16.) \qquad \Psi(z) = \Psi_{up}(z)|up\rangle + \Psi_{down}(z)|down\rangle$$

Here $|up\rangle$, $|down\rangle$ are states in spin space. During the passage Ψ_{up} gets deflected upwards and Ψ_{down} gets deflected downwards. And we

---

[i]Bell, op. cit., pp. 130—132.

[j]David Bohm, *Quantum Theory*, Prentice Hall, New York (1951). See especially Chapter 22.

obtain approximately

(17.) $\Psi(z) = \Psi_{up}(z - \delta)|up\rangle + \Psi_{down}(z + \delta)|down\rangle.$

If δ is greater than the width of the initial wave packet, then the up-part and the down-part have become separated in space, and it remains to measure the particle's position in order to complete the experiment. Since the position is $|\Psi|^2$ distributed, the probability of finding it in the upper channel is $\int |\Psi_{up}(z)|^2 dz$, which is the usual value. Thus in Bohmian mechanics there are two possible types of trajectories for the particle and one or the other is selected with a probability given by the quantum probability. This is also a nice illustration of the fundamental role that position measurements play. Measuring something as arcane as a spin comes down to measuring the position of a particle.

I may have explained enough to give you a sense of Bohmian mechanics and perhaps have left you with a desire to learn more. Even if you never apply Bohmian mechanics to anything, learning about it will I think gives a greater understanding of quantum theory.

Appendix A: The Derivation of Newton's Law[k]

In this appendix I want to present the steps that take you from the quantum version of the Hamilton–Jacobi equation to the quantum version of Newton's law; i.e. from Eq. (5) and $dQ/dt = \nabla S/m$ to Eq. (7) of the text. It will become clear from the derivation that these two representations of the mechanics are equivalent. One can choose either one depending on convenience. In the examples I have given in the text I have generally chosen Newton's law since I think this illuminates the physics. It requires introducing the quantum potential which is how the quantum mechanics enters. I begin with the quantum Hamilton–Jacobi equation,

(A1.) $\dfrac{\partial}{\partial t}S + \dfrac{1}{2m}(\nabla S)^2 + U + V = 0,$

---

[k]I am pleased to acknowledge the help of Basil Hiley in preparing this appendix.

where V is the ordinary potential and U is the quantum potential as defined in the text. Taking the gradient we find

(A2.) $$\frac{\partial}{\partial t}\nabla S + \frac{1}{2m}\nabla(\nabla S)^2 + \nabla(U + V) = 0.$$

We may now employ the identity for any vector A

(A3.) $$1/2(\nabla A^2) = A \cdot \nabla A + A \times (\nabla \times A),$$

and find, using the fact that A is a gradient so that the curl term vanishes,

(A4.) $$\left(\frac{\partial}{\partial t} + 1/m\nabla S \cdot \nabla\right)\nabla S + \nabla(U + V) = 0.$$

If we identify $\nabla S/m$ with the velocity v then the first expression in parenthesis is its total time derivative. Hence Eq. (4) is Newton's law,

(A5.) $$\frac{dp}{dt} = -\nabla(U + V). \qquad \text{QED}$$

Appendix B: More Harmonic Oscillator

This appendix was inspired by a reading of the paper "Bohmian mechanics as a heuristic device: Wave packets in the harmonic oscillator" by Gary E. Bowman.[1] I am going to make a very simplified version of this discussion. Historically speaking after Schrödinger invented wave mechanics he tried very hard to make it into a classical theory. The wave packets describing the simple harmonic oscillator seemed like a good candidate. It is well known that the ground state of the oscillator has the minimum possible uncertainty $\Delta x \Delta p = \hbar/2$. This suggests that wave packets associated with this state might show some classical behavior. This is best exhibited by examining the so-called coherent states. These are states with a Gaussian form where the maximum of the Gaussian obeys the equation of motion of a

---

[1]Gary E. Bowman, *AJP* **70**, (2002) 313–315.

classical oscillator. Explicitly

(B1.)
$$\psi(x,t) = \alpha^{1/2}/\pi^{1/4} \exp[-1/2(\xi - \xi_0 \cos(\omega_c t))^2$$
$$- i(1/2\omega_c t + \xi\xi_0 \sin(\omega_c t - 1/4\xi_0^2 \sin(2\omega_c t))].$$

Here $\xi = ax, \xi_0 = \alpha a, \alpha = (m\omega_c/h)^{1/2}$. The frequency ω_c is given in terms of the oscillator constant k by the relation $k = m\omega_c^2$. This somewhat complicated looking function has a Fock space interpretation as the eigenfunction of the annihilation operator. Explicitly the normalized eigenstate of the annihilation operator a corresponding to the eigenvalue α is given by $|\alpha\rangle = \exp(-|\alpha|^2/2)\exp(\alpha a^\dagger)|0\rangle$. Here $a^\dagger$ is the Fock creation operator and $|0\rangle$ is the Fock vacuum. This is what is meant by a coherent state. We can use Eq. B1 to find $|\Psi(x,t)|^2$. Thus

(B2.) $\qquad |\Psi(x,t)|^2 = \alpha/\pi^{1/2} \exp[-\alpha^2(x - a\cos(\omega_c t)^2)].$

This shows how the maximum of the wave packet follows the classical equation of motion. None of this has anything to do with Bohmian mechanics. But we can use Eq. B1 to compute the quantum potential U(x). Thus

(B3.) $\qquad U(x) = -m\omega_c^2/2(x - a\cos(\omega_c t))^2 + \hbar\omega_c/2$

If we add this to the classical potential then the term quadratic in x cancels out. "Newton's law" reads

(B4.) $\qquad dp/dt = -m\omega_c^2 a\cos(\omega_c t).$

Hence the Bohmian particles undergo classical oscillations.

Appendix C: Time Reversal

The question raised here is: is Bohmian mechanics invariant under time reversal? This must be answered in two parts. In the first place there is the Schrödinger equation that determines $\Psi(x,t)$. It is well-known that, providing the Hamiltonian does not depend explicitly on time, $\Psi(x,-t)^\dagger$ obeys the same Schrodinger equation as $\Psi(x,t)$ which is the way time reversal expresses itself in ordinary quantum mechanics. What about Newton's law which determines the behavior in time of Q? Clearly $d^2/dt^2 = d^2/d(-t)^2$ which assures us of the invariance of the left hand side of Eq. (7). That leaves the right hand side. Here our concern is the quantum potential. But this depends on $\Psi(x,t)^\dagger \Psi(x,t)$, see Eq. (6). Under the same transformation that leaves the Schrödinger equation invariant and this goes into $\Psi(x,-t)^\dagger \Psi(x,-t)$. This insures the time reversal invariance of the theory. Note that in the previous appendix the quantum potential was invariant under the exchange of t with −t, giving an example. See Eq. (B3).

Acknowledgements: I am grateful to Elihu Abrahams, Steve Adler, Freeman Dyson, Arthur Fine, Jim Hartle, Basil Hiley, and Peter Kaus for helpful communications on this subject, as well as an anonymous referee for a very careful reading of this essay. The spirit of John Bell hovers over what I have written. As a parallel, in gratitude for his help T. S. Eliot dedicated his poem *The Waste Land* to Ezra Pound. The dedication reads "Il miglior fabbro..." — the better craftsman.

7. Max Born and the Quantum Theory

Reprinted with permission from *Am. J. Phys.* **73**, 999 (2005).
Copyright American Association of Physics Teachers

1. Introduction

Of all the founders of the quantum theory, the least known to
the general public is the German-born theoretical physicist Max
Born. Einstein, Bohr, Schrödinger, Dirac, and Heisenberg have been
the subjects of full-scale biographies[1] and in the case of Bohr and
Heisenberg even a play.[2] Until very recently there was no biography of Born. Now there is *The End of a Certain World* by Nancy
Thorndike Greenspan.[3] Although this essay is not primarily a book
review, I would like to note that her grasp of the science is shaky.
Born published various autobiographical memoirs in which he aired
his grievances about what he considered his lack of recognition
for the part he played in the creation of the quantum theory. It
took until 1954 before he was awarded the Nobel Prize for work
that he had begun in 1925. Even then, the prize, as I will later
explain, was given for only a small fraction of what he and his collaborators had done. Born also resented some of the people with
whom he had worked. He believed that Heisenberg failed to make
clear the intellectual debt that he owed Born from the time when
he was Born's assistant in Göttingen, and he thoroughly disliked

Robert Oppenheimer who had been his student. He once described Oppenheimer as someone who oscillated between "arrogance and a somewhat unpleasant modesty,"[4] as far as I am concerned, not a bad description. The feelings must have been somewhat mutual because when Oppenheimer delivered the Reith Lectures on the BBC in 1953, he omitted mentioning Born when he described the history of the creation of the quantum theory. This omission provoked a letter from Born and a response from Oppenheimer who said that he had kept the names to a minimum to avoid confusion. He then added, "I am one of the last to forget your part in these affairs, for I was one of those who learned of the discoveries more or less as you made them."[5] Oppenheimer was good at writing letters like this.

Greenspan, with the help of the surviving Born family, which includes Born's granddaughter, the British-born pop-singer Olivia Newton-John, has provided an account of Born's life that is so detailed that one sometimes has the feeling that it is taking place in real time. All of this detail is admirable and makes for excellent biographical material. What is less admirable is Greenspan's attempt to explain the science which she does not understand. I will come back to the science shortly and also to an account of Born's life. But first I want to comment on the biographies of scientists written by biographers who do not understand their work. Greenspan's biography is the second such biography I have read recently. The other was *Obsessive Genius: The Inner World of Marie Curie* by Barbara Goldsmith.[6] The books have in common the use of material that was made available by the families of the subject. In Goldsmith's case it was Madame Curie's granddaughter Hélène Langevin-Joliot who helped Goldsmith have access to this material. In both cases there were reputable physicists who seem to have vetted the manuscripts. Nonetheless, in Goldsmith's book there are too many descriptions such as the following which has to do with Lise Meitner's identification of fission, "Also, she knew the exact mass of uranium and barium,[7] and realized that some of this mass disappeared in the splitting process. Meitner calculated that one gram of uranium contained the unbelievably huge number of 25 followed by 20 zeros. By using Einstein's formula of relativity in her calculations, she found that the energy discharged when the mass disappeared and that when the atom split were roughly the same."[8] When I was teaching, I used to have students who wrote similar passages. Did none of her editors or consulting physicists ask the question "unbelievably huge number of what?" Did no one ask what "roughly the same" means? Why "roughly"? Why not "exactly"? Is she implying that energy is not conserved? I heard Goldsmith being interviewed on television about her book. She said that she thought the science was "easy." It certainly was easy compared to the physics that Greenspan had to discuss. During the interview a member of the audience noted another mistake to which Goldsmith responded that she was aware of it, but did not want to be more detailed in her book for fear of overburdening her readers. That is, she can lighten their burden by producing incorrect science.

Having scientists vetting similar books is probably necessary, but it is not sufficient. It depends on how the author chooses to use the information provided. In 1980, I was asked by a publisher to vet a manuscript of C. P. Snow. It was published the next year under the title *The Physicists*. It was an account of 20th century physics including the quantum theory. I was surprised by how little Snow seemed to understand. I guess by that time he had not done any physics for half a century. I was a great admirer of some of his novels although less so of his "two-cultures." So I decided to put my shoulder to the wheel and sort out the mess. I thought that perhaps Snow would send me a note thanking me for saving him from making a fool of himself. Not at all. The publisher told me that Snow was furious and was not going to change anything. Someone must have persuaded him to do so because the final book does not have most of the mistakes that I had noted.

2. Summary of Born's Life

Max Born was born—the alliteration seems unavoidable — in Breslau, now Wroclaw, Poland on December 11, 1882. His father Gustav was a research biologist who was not able to advance to a significant university position in large part because he was Jewish. He was told that he might advance if he converted to Lutheranism which he refused to do. Later Born did convert, largely to avoid conflicts with his in-laws who were converted Jews. He wrote, "It has not changed me, yet I never regretted it. I did not want to live in a Jewish world, and one cannot live in a Christian world as an outsider. However, I made up my mind never to conceal my Jewish origin."[9] Born senior married into the wealthy and thoroughly assimilated Kauffmann family who disapproved of his modest position and status. In 1884, Born's sister Kathe was born and 2 years later their mother died. It is plausible that Born's sense of insecurity and neuroticism throughout his life (He had mini-nervous breakdowns and suffered from illnesses that one might think were in part psychosomatic.) had much to do with the loss of his mother. When he was young, Born was considered too fragile to go to school and was tutored at home by a man whose last name was, remarkably, Böhr. In 1892, Born's father remarried. His second wife, Bertha Lipstein, came with a substantial dowry and also, it appears, an eastern European Jewish accent that made her unacceptable to his previous in-laws. She seems to have been a kind woman who looked after Born and his sister. But in 1900, Born's father died from a heart attack at age 49. This turmoil in his family contributed to Born's almost pathological shyness. By this time he was finishing his secondary education at the Kaiser Wilhelm Gymnasium in Breslau. He then matriculated to the university in Breslau. Because he was financially secure, he was not concerned about training for a profession and wandered among different academic disciplines such as philosophy and astronomy until he discovered higher mathematics. It was fortunate for him that a professor of mathematics named Jakob Rosanes was at the university. Rosanes did fine work in fields such as algebraic geometry and was also well-known as a chess player. His games are still the subject of active discussion and analysis. Of more importance to us is that he taught Born about the algebra of matrices, which became crucial in Born's contribution to the quantum theory. In the after dinner speech that Born gave at the Nobel Prize banquet, he acknowledged his debt to Professor Rosanes.

With his new interest in mathematics, which was to play a dominant role in Born's work and later his career, it was almost inevitable that he would continue his studies in Göttingen. The latter was then arguably the mathematical capital of the world, a reputation that it had held for decades.[10] I once read that if you went into a tavern in Göttingen with a new and elegant proof of a theorem, you got free drinks from the tavern keeper. I have no idea if this was true, but at the time Born arrived there, there were three great mathematicians on the faculty: Hermann Minkowski, Felix Klein, and David Hilbert, who is regarded by many people as the greatest mathematician of the 20th century. Greenspan notes that every Thursday afternoon these mandarins took a traditional walk on which very occasionally junior colleagues were allowed to go along. Born must have already shown signs of great promise because he soon became Hilbert's "scribe." Hilbert did not lecture from a text, and it was the job of the scribe to take very good notes for the benefit of the other students. It also meant spending time with Hilbert and eventually joining the mandarin's walk.

All three of the mandarins had a deep interest in physics. Hilbert once famously remarked that physics "was much too hard for physicists."[11] In 1904–1905 Klein organized a seminar on elasticity to which Born made a significant original contribution. Klein expected him to continue working this subject for his dissertation and when Born had other ideas, Klein was furious. But Hilbert, Minkowski, and some junior faculty held a seminar on electrodynamics. This seminar was the first time the students were exposed to the work of Lorentz and Poincaré. They did not study Einstein's theory at this time, but both Minkowski and Hilbert made monumental contributions to it later. Minkowski's four-dimensional space-time is the one we teach and use.

In the end, Born did complete his Ph.D with Klein in 1906 on elasticity, and immediately was conscripted into the army for his compulsory one year tour of duty. He was discharged after a short service because of an asthma attack, something that would re-occur throughout his life especially when he was under stress. This discharge was followed by a brief stint at Cambridge before he returned to Breslau. At this point he had to make a decision. Having been around truly great mathematicians, he had come to the realization that he was never going to be one of them. He decided that the next best thing might be physics. After some looking around for someone to work with, he received an invitation from Minkowski to come to Göttingen to collaborate on some problems connected with relativity. This invitation was for the fall of 1908, and by the next January, Minkowski had died because of a ruptured appendix. He was 44. By this time Born had gone far enough with the research so that he could carry on alone. The work was good enough so that he was invited to become a lecturer in Göttingen. One of the nice anecdotes Greenspan tells us is that, in Göttingen, Born shared a house with three men and a woman, a pretty black-haired medical student named Ella Philipson. She subsequently married a physicist Paul Peter Ewald who had often visited the house. She does not tell us that Hans Bethe married their daughter Rose in 1939. Bethe had been one of Ewald's assistants at the University of Stuttgart. The physics world is small indeed. In 1912, Born met Hedwig Ehrenberg, "Hedi" as she was known to her friends. After a year they were married. Much of Greenspan's book describes their troubled relationship which endured until Born's death in 1970. They often lived apart, each one nursing their own neuroses. They were unfaithful to each other. Hedi had a long relationship with a Göttingen mathematician named Gustav Herglotz whom she thought of marrying. She once told her daughter Irene that she never really loved Born and only married him because he seemed so eager to get married. If the two them had not been forced out of Germany by the Nazis and thrown together for survival, it seems unlikely that the marriage would have lasted.

The relationship between the Borns and Einstein deserves a whole book by itself. It lasted 40 years and produced a wonderful collection of letters which has recently been republished.[12] It began in 1916, when Einstein paid his first visit to the Born's home in Berlin. Born had accepted a sort of assistant professorship in Berlin which was a promotion from his Göttingen lectureship. Apart from physics, what drew the two of them together was music. Born was an accomplished pianist, and Einstein, as is well-known, played the violin. It is interesting how many of these European physicists were also musicians. Max Planck at one point had to decide whether he was going to be a concert pianist or a physicist and Werner Heisenberg was an excellent pianist. The year when Born and Einstein got to know each other was also a year in which the First World War was raging. Born, like many German Jews, was a very patriotic German. He was once again called into the army and, at least in the beginning, applauded any successes the Germans had in battle. It is to his credit that he refused an offer from another German Jew, the chemist Nobelist Fritz Haber, to work on poison gas. Einstein was an outspoken pacifist, which did not endear him to his German colleagues. He could not have cared less. The Borns regarded Einstein as an island of sanity in both the chaos in their own lives and in what was going on around them. Hedi Born commented on Einstein's laugh. The late Abraham Pais remarked that it sounded like the barking of a seal. I had this characterization in the draft of the New Yorker profile I wrote of Einstein. I had it vetted by Helen Dukas, his long-time secretary. Miss Dukas objected to this description, but did agree that he had a very hearty laugh. I kept the barking seal in.

In 1921, after an interval in Frankfurt, Born accepted a position at Göttingen — this time as a full professor which in the German academic system meant the head of his own department. He managed to negotiate positions for assistants. One of the first was Pauli who did not stay long because he did not like life in a provincial town and because he thought Born's approach to physics was too formal. Born was, after all, first trained as a mathematician. Heisenberg, who did not yet have his Ph.D, succeeded him. This assistantship was a baroque arrangement because Heisenberg was nominally a student of Arnold Sommerfeld in Munich. But Sommerfeld was lecturing in America and agreed to loan Heisenberg to Born. It was already recognized that Heisenberg was special. By 1923, Born had shifted his interests to the quantum theory—the "old" quantum theory. The new quantum theory began with Heisenberg the following year.

Writing about Born is very different than, say writing about Einstein. With Einstein it is a given that he was one of the greatest physicists who ever lived. It is quite possible to write books about aspects of his life in which his physics is barely relevant. With Born it is quite different. He did his best, indeed his only really great work, in this relatively brief period in the 1920s. Absent that, it is difficult to see why anyone would devote a full-scale biography to him. Thus understanding his work is pivotal to a biography. I am going to try to make an account of Born's work.

3. Born's Science

By the end of the 19th century, the experimental study of atomic spectra was well underway. A collection of atoms at a finite temperature gives off radiation, some of it visible. If this radiation is put through a spectrometer, a series of lines are produced. The wave lengths, intensities, and spacing of these lines characterize the atom. They are a kind of fingerprint. From these fingerprints the elements in the Sun and other stars can be identified for example. Explaining why atoms had these characteristic spectra was a problem for theoretical physicists which became acute when Ernest Rutherford published in 1911 his explanation for some experiments that had been done by his young collaborators Hans Geiger and Ernest Marsden. They had been using alpha particles, helium nuclei, to scatter off a gold foil. In a few of these collisions the alpha particles were scattered at large angles, which Rutherford correctly explained by postulating that they had hit something hard in the gold atom. Thus the atomic nucleus was discovered. The model that emerged of the atom was that of a massive, positively charged nucleus surrounded by a cloud of much less massive electrons. The simplest case, which will suffice for our purposes, is that of hydrogen which consists of one massive proton nucleus and one orbiting electron. It became clear that this model could never explain atomic spectra if classical physics was invoked. Here is the reason.

In 1897, the Irish physicist Joseph Larmor presented a formula based on classical electrodynamics for the energy per second that an accelerating electric charge would radiate. It turns out that the energy radiated per second is proportional to the square of the acceleration of the electric charge. Let us consider a simple case. More complex cases can be inferred from this example. Suppose the charge oscillates with a frequency ω. We can replace the square of the acceleration by the square of the position of the electron, multiplied by the fourth power of the frequency, ω^4. This relation might appear to offer a way of explaining the spectral lines. But as the electron orbits around the proton, it radiates electromagnetic energy continuously. But classical physics tells us that as it does so, it moves closer to the nucleus. Indeed, in short order it will collapse into the nucleus. So as a consequence we need to account for atomic stability. In addition, the frequencies that an atom would emit during its orbit would resemble nothing like the beautiful spectral lines that are observed. An image that often occurs to me is pushing a grand piano off of a roof and expecting it to play Beethoven's Moonlight Sonata upon impact. Enter now Niels Bohr.

In 1912, Bohr who was in his late twenties, went to Manchester to study with Rutherford. He was therefore in an excellent position to learn about the new atomic model. However, he did not do anything about atomic spectroscopy until he returned to Denmark that year. It was at this time that he learned of some work that had been published by a Swiss high school teacher, Jakob Balmer, in 1885, which was, coincidentally the year of Bohr's birth. Balmer, who had a Ph.D. in mathematics, found a simple formula that fit the very limited data that was then available on the hydrogen spectrum. We would write his result for the frequencies ω_{nm} as $\omega_{nm} = C(1/n^2 - 1/m^2)$, where C is a constant and n and m are positive integers with m greater than n so that the frequency is positive. To fit the observations Balmer took n = 2 so that m began with 3. This formula defines what is called the Balmer series. Other series are possible with different choices of n. Balmer's remarkable result provided the essential clue for Bohr who made the following leap. He supposed that instead of allowing the electron to have all possible orbits around the proton, only select ones were allowed. We now call these Bohr orbits. Furthermore, he supposed that the energy in the nth orbit is proportional to $-1/n^2$. The minus sign appears because the electron is bound to the proton. Furthermore, he assumed that radiation was emitted only when an electron jumped from a higher energy orbit to a lower energy orbit. The frequency of the emitted energy was then proportional to the difference of energies. Thus in his model we have $\omega_{nm} = 1/\hbar$ (Em − En), where $\hbar$ is the quantum constant introduced by Planck in his early papers on black body radiation. Bohr was also able to derive the constant of proportionality for the energy and found that it agreed with experiment. Bohr's model was certainly progress.

Before I go on, let me quote how Greenspan describes this theory. She writes, "The frequency of the electron is proportional to the difference in energy between the two levels."[13] Reading something like this is like hearing fingernails being scratched on a blackboard. If an author who was tone deaf were to propose writing about the music of Bach or one who was color-blind propose to write about the art of Picasso, a publisher might object. But publishers who publish biographies of scientists written by non-scientists with no real understanding of the science seem to have no problem. It is a source of constant amazement to me. I go on with the science.

Bohr's energy level picture appeared to give a good account of the position of the spectral lines — at least for simple atoms like hydrogen. But there was another matter to be explained, the relative intensity or brightness of the various lines. The classical physics explanation was via the Larmor radiation formula. According to this picture a single electron would continue radiating as long as it was accelerating. It could by itself produce a spectral line. But the Bohr quantum picture was entirely different. When an electron makes a quantum jump from one energy level to a lower one, it emits a single quantum in general. One quantum does not a spectral line make. It requires myriads. These come about because in this picture there is an ensemble of atoms each one contributing its photons. What then accounts for the relative intensity of the spectral lines? The reason, according to this picture, is that there are different relative probabilities associated with different transitions. When the probability of a quantum jump is large, the associated spectral line will be bright. The idea of introducing probability amplitudes to describe such radiation processes dates back to three papers that Einstein wrote in 1916–1917.[14] In one of them Einstein introduced the physics that decades later became the basis of the laser. Trying to estimate these probability amplitudes for transitions in atoms like hydrogen was actively carried out by practitioners of the "old" quantum theory with mixed success. The main tool that was being used was what Bohr called the "correspondence principle." Bohr always took the position that there were limits in which quantum physics merged with classical physics. For spectra like that of hydrogen, it was the limit where the energies became large, which was the limit in which the quantum number became large. In this limit the energy levels are so close together that, in effect, they form a continuum just as they would in classical physics. One can readily show that in this regime the energy differences, compared to some fundamental energy, have the pattern of a classical oscillator. Thus one can apply the Larmor formula in this regime to compute intensities and then hope that it worked elsewhere in the spectrum.[15] This reasoning was hardly a very satisfactory way to construct a fundamental theory, but people were stuck. Enter Heisenberg.

People may argue—Born certainly did—about who deserves credit for certain aspects of the development of the quantum theory. But of one thing there can be no argument, that is, the importance of the work done by Heisenberg and Schrödinger. What they did was unique. Of the two I think Heisenberg deserves the most credit. He had no predecessors. The rest built on him. In early 1925, Heisenberg was in Copenhagen on a temporary leave from Göttingen trying to understand some of the problems with the old quantum theory. When Heisenberg returned to Göttingen in June, he suffered from an extreme attack of hay fever. He went to a tiny resort island called Helgoland in the North Sea to seek relief. By the time he returned to Göttingen, he had discovered the rudiments of quantum mechanics. His seminal 15 page paper with the innocuous title "On a quantum-theoretical interpretation of kinematic and mechanical relations," was published in the September, 1925 issue of Zeitschrift für Physik.[16] He had completed the paper in early July and had sent one copy to Pauli and had given the other to Born to read in order to solicit his advice as to whether or not to publish it. I will describe Born's reaction after I outline what Heisenberg did.

This paper is notoriously difficult to read.[17] Part of the reason is that Heisenberg did not completely understand it himself. He had invented what appeared to be a new mathematics without realizing that it was a form of mathematics well-known to mathematicians but not to physicists. This was matrix algebra, as Born rapidly discovered. But he also made leaps in his reasoning that could not really be justified, which is hardly surprising. Quantum mechanics cannot be derived from classical physics. Heisenberg began his paper by stating his philosophy. He writes, "It is well known that the formal rules which are used in quantum theory [the "old" quantum theory] for calculating observable quantities such as the energy of the hydrogen atom may be seriously criticized on the grounds that they contain as basic element [sic], relationships between quantities that are apparently unobservable in principle, e.g., position and period of revolution of the electron. Thus these rules lack an evident physical foundation, unless one wants to retain the hope that the hitherto unobservable quantities may later come within the realm of experimental determination. This hope might be justified if the above-mentioned rules were internally consistent and applicable to a clearly defined range of quantum mechanical problems."[18] Instead the new mechanics should deal only with relations among observables. The electron's Bohr orbits are not observable, but the pattern of the spectral lines and their relative intensities are. In the classical Larmor formula for radiation from an oscillating charge, the power radiated is proportional to the square of the electron's position as a function of time. But this position is not an observable, and so in the quantum theory it should be replaced by the square of the probability amplitude, A_{nm}, for the electron to make a transition from a state with energy E_n to a state with energy E_m.[19] In the quantum theory, this transition need not be direct. The electron can transition to some lower energy state from it and transition to the final state. Naively one might imagine accounting for this possibility by adding up all these transition amplitudes and then squaring the result. Heisenberg realized that something physically was wrong with this possibility.[20] It violated what was known as the Rydberg-Ritz combination principle, which really comes down to the conservation of energy. Suppose the electron makes a transition to its final state through some intermediate state with an energy, say, E_j. Then the total energy of the quanta released must be $E_n - E_m$. Whatever the sequence of quantum jumps, the electron must end up with this energy. If you just multiply the two transition amplitude sums together without restrictions, there will be cross terms that violate this principle. Instead you must restrict the terms in the multiplication so that a transition to an intermediate level with E_j is followed by a second transition from this level to the final one. In this case the net energy emitted will be equal to $E_n - E_j + E_j - E_m = E_n - E_m$. This reasoning tells us how we must do the quantum multiplication. If the combined amplitude, say Cnm, is obtained from the amplitudes Anj and Ajm we must have the multiplication rule, Cnm = ΣjAnjAjm. Note the index j that is summed over, is common to the factors in each term of the sum of the Anj's. This multiplication rule was the new quantum algebra.

Heisenberg made a discovery about this strange algebra. Suppose we have two quantities Anj and Bjm. We can combine these as before to compute Cnm = ΣjAnjBjm. But what happens if we combine them in the other order, C′nm = ΣjBnjAjm? Are C and C′ the same? In general the answer is no. Symbolically we can write AB ≠ BA. In this algebra quantities do not commute. In his paper Heisenberg notes, "A significant difficulty arises, however, if we consider two quantities X(t), Y(t) and ask after their product X(t)Y(t) ... Whereas in classical theory X(t)Y(t) is always equal to Y(t)X(t), this is not necessarily the case in quantum theory."[21] This "significant difficulty" turned out to be the heart and soul of quantum mechanics.

After Heisenberg gave Born his paper to read, he went off on vacation. Born studied it and very soon realized that he knew what this strange algebra was. Heisenberg had invented for himself the notion of "matrix" multiplication, which Born had learned many years earlier from Professor Rosanes in Breslau. To a contemporary physicist the idea that Heisenberg had never heard of matrices seems almost incredible. Mathematicians had been using them in Europe for centuries to solve simultaneous linear equations and their use in the Orient is even older. The name was invented in 1850 by the English mathematician James Sylvester who worked on them. But until the development of quantum mechanics, they had rarely been used by physicists.[22] Born had used them in a paper he wrote on relativity in 1909 and in 1908 Minkowski had written a paper that had a pedagogical discussion of matrix algebra.

If all Born had done was to expose a limitation in Heisenberg's mathematical erudition, it would hardly be worth noting. But the next step was extremely important. Heisenberg produced a formula involving the non-commutivity of the probability amplitudes. I used the word "produced" because it cannot be derived from anything in classical physics. Its production was part of Heisenberg's magic. The formula is so important that I will write it down even though it will look a little ungainly. All the elements will be familiar to us. The formula reads with a notation slightly different from the one that Heisenberg uses,

$$(1.) \quad h = 4\pi m \sum_{j=0}^{j=\infty} A_{nn+j}A_{n+jn}\omega_{n,n+j} - A_{nn-j}A_{n-jn}\omega_{n,n-j}.$$

Here the ω's s are proportional to the energy differences. For example $\omega_{n,n+j} = 2\pi/h(E_n - E_{n+j})$, h is Planck's constant, the As are probability amplitudes, and m is the mass of the electron. The sum uses Heisenberg's matrix multiplication. I apologize for the ungainly sight of Eq. (1). I give it here so that you can appreciate what Born did with it.

I will not try to carry out the details.[23] Here is the outline. We have seen that by using the correspondence principle, the classical position x(t) becomes the probability amplitude A_{mn}. Let us rename this $x(t)_{nm}$. In other words, the quantum mechanical position is represented by a matrix. We saw that for the oscillator the acceleration became $\omega_{nm}^2 A_{nm}$. Born claimed that the speed v would be represented by $i\omega_{nm}A_{nm}$. The appearance of i, the square root of -1, enters when you work out the details and cannot be avoided.[24] Thus we can rewrite the Heisenberg relation, after doing the sum carefully, as

$$(2.) \quad \sum(x(t)_{nj}V(t)_{jn} - V(t)_{nj}x(t)_{jn})m = ih/2\pi$$

If we recall that mV is the momentum, p, we can symbolically write Eq. (2.) as

$$(3.) \quad (xp - px)_{nn} = ih/2\pi$$

Equation (3.) gives the diagonal terms in the matrix. Born conjectured, correctly as it turned out, that the other matrix elements are all zero. If we define I as the unit matrix which has all ones on the diagonal and zeros elsewhere, we can write Eq. (3) as

$$(4.) \quad xp - px = ih/2\pi L$$

Equation (4.) is the canonical commutation relation of the quantum theory — canonical because of its fundamental importance. We make use of it when we derive Heisenberg's uncertainty principle which limits the simultaneous measurability of position and momentum in the quantum theory. It shows clearly how Planck's constant h sets the scale of these effects. If you set h equal to zero, you recover the classical physics answer. In addition, Born showed how these position and momentum matrices change in time when there are forces that come into play. At this point, for reasons I will now explain, the assignment of credit becomes somewhat murky.

At the time, Born experienced the mixture of physical and emotional exhaustion that frequently troubled him throughout his life. He went off for a rest cure. He turned over what he had done to a student named Pascual Jordan. Jordan was extraordinarily brilliant and was in the Heisenberg-Pauli league. What is not clear is which part of the work that he and Born published jointly was due to Jordan and which part was done by Born. My guess is that most of it was due to Jordan. This conjecture is consistent with letters that Heisenberg sent to Jordan at the time from Copenhagen.[25] In these letters Heisenberg discusses results that Jordan had obtained. One thing has always struck me as odd. The work that Born and Jordan was doing was an adumbration of work that Heisenberg had shared with Born. It would seem natural to me for Born to have invited Heisenberg in on this collaboration. Heisenberg does not seem at all troubled by this non-invitation to collaborate, and he wishes Jordan success in what he is doing. In any case, this work was published in the fall of 1925 in the *Zeitschrift für Physik*.[26] It begins with an acknowledgement to Heisenberg, "The theoretical approach of Heisenberg recently published in this Journal, which aimed at setting up a new kinematical and mechanical foundation in conformity with the basic requirements of quantum mechanics appears to us of considerable potential significance. It represents an attempt to render justice to the new facts by setting up a new and really suitable conceptual system instead of adapting the ordinary conceptions in a more or less artificial and forced manner. The physical reasoning which led Heisenberg to this development has been so clearly described by him that any supplementary remarks appear superfluous."[27] Having set the stage, they then proceed to explain the elements of matrix algebra and to write down the canonical commutation law. But they also do the matrix dynamics — how things change in time — including deriving an equation that is often, and I think incorrectly, ascribed to Heisenberg. This paper lays out the essence of what came to be called "matrix mechanics." It is all there.

This sequence of events raises a dilemma. Who should get the credit? To see why the answer to this question mattered, Einstein nominated Heisenberg, Born, and Jordan for the Nobel Prize in 1928, although he said that Heisenberg had the strongest case. As it happened, and on the matter of Born and the Nobel Prize, Greenspan is very helpful, although I don't think she does full justice to Jordan's somewhat unsavory story, there was on the five person committee of the Royal Swedish Academy, which awards the Nobel prize in physics, a man named Carl Oseen who was an old friend of Born. Despite this friendship, Oseen decided that the quantum theory, at least at that time, was too speculative to meet the terms of Nobel's will. A few years earlier the Academy had decided that relativity was too speculative to qualify for a Nobel Prize. Einstein was given it for the photo-electric effect. After 1928, Born and Jordan were mentioned occasionally in Nobel Prize nominations, but usually compared somewhat unfavorably to Heisenberg and to Dirac and Schrödinger. In 1932 Heisenberg was awarded the prize and in 1933 it was shared by Dirac and Schrödinger. Born and Jordan had dropped out of consideration. I will come to Born later, but let me say something about Jordan.

In 1926, Jordan took his degree with Born and soon afterward moved to Hamburg University until 1929 where he remained until 1944 when he went to Rostock University, finishing his career at Hamburg. As far as I can tell, none of the work that he did after leaving Göttingen compared to what he did while he was there. One of the things that he did while still at Göttingen, for which he did not get credit, was the invention of what are called Fermi-Dirac statistics, the statistics that are obeyed by ensembles of particles such as electrons. The reason he did not get credit is rather tragic. In 1925, Born made a trip to the United States to give various lectures. Before leaving, Jordan gave him a manuscript of the paper he had written on the statistics. Born put it in a suitcase and forgot to read it. By the time he did, it was too late. Fermi and Dirac had already published.

By the early 1930's Jordan seems to have become interested in subjects like the quantum theory and free will. He had what appears to have been a brief correspondence with Carl Jung. He also joined the Nazi Party on May 1, 1933 and became a Storm Trooper.[28] He later tried to claim that he had done so to try to keep German science from being totally Nazified. He was accused by more extreme Nazis of being a "White Jew," because he had had scientific collaborations with Born and worked on theories that had been invented by Jews. But, the fact is that he became a spokesman for the Party. In 1941, he wrote things like "... National Socialism will not be content with harnessing German scientific research; on the contrary, in appreciation of the fundamental importance of this research work for the life of the nation, it will provide German science with impetuses from its own energy and conception, out of which the dawn of a new great epoch of development can be expected..."[29] During the war he worked in Penemunde and elsewhere.

After the war Jordan tried to persuade Born to write a testimonial that would help him get a promotion at Hamburg. Former Nazi party members needed such recommendations above and beyond the usual ones. Born refused. In his letter of refusal he listed by name the members of his family who had been murdered by the Nazis. Heisenberg did provide such a document. My guess is that once the Swedish academy learned about Jordan's Party affiliations — which were no secret — that was the end of his chances to receive a Nobel. Because Born and Jordan were linked, it also, I believe, affected Born's chances. In any event, there were new rising stars, Dirac and Schrödinger.

When Dirac was awarded the Nobel Prize in 1933, a year after Heisenberg, Born was devastated. In re-reading these early quantum theory papers I think he has a case. It is true that Dirac's method is somewhat more general, but the application to the canonical commutation relations, which I think is the most important result. was first done and first published by Born. In this calculus Born always gave Jordan rather little credit. He said that the ideas were his and that Jordan just turned the crank.

With all the participants now gone — Jordan died in 1980, Born in 1970, Heisenberg in 1976, and Dirac in 1984 — we will probably never be able to sort this out. It is often said that Dirac did so much after this first paper, in particular the Dirac equation in 1928, which for the first time connected quantum mechanics and relativity and predicted anti-matter, that he deserved a Nobel Prize under any circumstances. But keep in mind that Born was eventually awarded his prize for work, which I will soon discuss, done after his first quantum papers with Jordan. Moreover, as late as 1933, the Dirac equation was still regarded with some skepticism. You need only read Pauli's famous 1933 monograph on quantum physics to see the sort of objections that there were raised; in particular, the negative energy solutions were thought to be unphysical.

Born reacted, I suppose predictably, to Heisenberg's Nobel Prize and was very upset by it. Someone with a stronger ego might have found satisfaction in the success of a protégé. But not Born. Remarkably, Heisenberg seems to have shared his feelings. Heisenberg wrote to Born in 1933, when he won the prize, officially the 1932 prize, which had been delayed. By that time Born had left Germany and was in Cambridge. Heisenberg posted the letter from Switzerland so that it would not be opened by a censor. He wrote, "Dear Herr Born, if I have not written to you for such a long time, and have not thanked you for your congratulations, it was partly because of my rather bad conscience with respect to you. The fact that I am to receive the Nobel Prize alone, for work done in Göttingen in collaboration — you Jordan and I (This is a reference to a long paper published in 1926, often referred to as the *Dreimannerarbeit* — the 'three-man-work.' It spells out in detail matrix mechanics. It is a remarkable and very profound paper. Again it is unclear which author contributed what.) the fact depresses me and I hardly know what to write you. I am, of course, glad that our common efforts are now appreciated, and I enjoy the recollection of the beautiful time of collaboration. I also believe that all good physicists know how great was your and Jordan's contribution to the structure of quantum mechanics — and this remains unchanged by a wrong decision from outside. Yet I myself can do nothing but thank you again for all the fine collaboration and feel a little ashamed. With kind regards. Yours, W. Heisenberg."[32]

Although I cannot help admiring the impulse that led Heisenberg to write this letter, in retrospect, it seems to me to be a little overdone. As I have tried to emphasize, Heisenberg's contribution was unique. He broke the logjam of classical physics, something I am persuaded that Born, gifted as he was, could never have done. He did not have that kind of mind. Dirac also felt that Heisenberg played a unique role. I know, because he told me. When I was writing profiles of physicists for the New Yorker in the 1960's, I interviewed Dirac with the intention of writing a profile of him. He was again visiting Princeton. Because Dirac did not feel like spending any more time, I had only one interview, and I abandoned the project. But in that interview Dirac emphasized the importance of the role of Heisenberg. I don't recall him mentioning Born. The other unique contribution was that of Schrödinger, to which I now turn.

As I have noted before, in 1924, the French aristocrat and physicist Louis de Broglie made the speculation in his doctoral thesis that particles like electrons had a wave nature. This property was the counterpart to light, which Einstein had argued in 1905 had a particle as well as a wave nature. The thesis was sent to Einstein who remarked that he had had ideas along these lines, although he had never spelled them out. In 1927, the Bell Telephone Laboratory physicists C. J. Davisson and L. H. Germer and independently, G. P. Thomson in England, showed that this wave nature was true experimentally. DeBroglie was awarded the Nobel Prize in 1929 and the three experimenters in 1937. The question was what was the significance of these waves? The first thought, one that Einstein liked, was that these were waves that oscillated in ordinary space and acted to somehow guide the electrons. In the first half of 1926, before the experiments had been published, the Austrian theorist Erwin Schrödinger, in a monumental series of papers, proposed the equation that these waves satisfied and proceeded to solve several problems with it. For a brief period it looked as if there were two quantum theories — matrix mechanics and wave mechanics. Einstein liked the wave mechanics, at least for a while, because it seemed much like classical physics. Heisenberg hated wave mechanics for the same reason.

This disagreement came to an end when Schrödinger showed that the two theories were equivalent. They were simply different representations. In matrix mechanics it is the properties of the matrices that are of interest and not the space on which these matrices operate. In the wave mechanics we study the properties of the elements of this space. For example, to solve the hydrogen atom, the matrix people studied the structure of the energy matrix, and the wave people studied what the solutions to the wave equation were when you invoked this matrix on the space of possible solutions. If you started with the matrix equation, you could derive the Schrödinger equation and vice versa. It was a matter of convenience which you used. In general, the wave equation was much more convenient.

The equivalence between the two formulations still left open what these waves meant. This meaning was settled by Born, and it was what eventually won him the Nobel Prize. Born applied the Schrödinger wave mechanics to the problem of quantum mechanical collisions. Typically one begins with two particles widely separated in space. It is assumed that the mutual forces that act on them essentially vanish when the particles are separated. When the particles approach, they interact causing them to separate; when the separation is sufficiently large, the particles are again free of interaction. But the interaction has altered their momenta so that the scattered particles are now found at various angular separations from each other. Born treated this problem using the Schrödinger wave mechanics. The approximation method he used — the "Born approximation" — is the one we routinely teach our students. It gives good answers if the forces are relatively weak.

In June of 1926, Born wrote the first of two papers on the subject.[33] In the first he stated the interpretation of the Schrödinger wave function that has been with us ever since. Let us suppose that this wave function is a function of position and time. He postulated that this function was the probability density for finding, say the electron, at a given position at that time. To find the probability you take the square of the function.[34] In his paper he stated his conclusion, "One does not get [in the quantum theory] an answer to this question, What is the state after collision? But only to the question, How probable is a given effect of the collision? ... From the standpoint of our quantum mechanics there is no quantity [*Grösze*] which causally fixes the effect of a collision in an individual event. Should we hope to discover such properties later ... and determine [them] in individual events? . . . I myself am inclined to renounce determinism in the atomic world, but that is a philosophical question for which physical arguments alone do not set standards."[35] These philosophical questions have been with us eversince.

The first dissenter was Einstein. He wrote Born an often quoted letter in which he said, "Quantum mechanics is very impressive. But an inner voice tells me that it is not yet the real thing. The theory produces a good deal but hardly brings us closer to the secret of the Old One. I am at all events convinced that He does not play dice. . ."[36]

For the rest of Einstein's life he and Born disagreed about this interpretation. The discussion was very civilized until nearly the end. Finally, Einstein lost his patience and would only discuss this issue with Born through an intermediary, Pauli, who spent the war at the Institute for Advanced Study. Most physicists believe, in so far as they think about the matter, that Einstein was wrong. When Born won the Nobel Prize in 1954 for this work, he seemed to have been somewhat surprised. The rest of what he had done was not noted. In his Nobel lecture he presented his version of the entire history.[37] If I had to hazard a guess — the proceedings of the Royal Swedish Academy, when they become available, are notoriously laconic — it would be that they avoided the rest of the work because they did not want to include Jordan who was still alive. The statistical interpretation of the quantum theory we owe to Born alone.

As far as I can judge, these quantum mechanics papers were the high water mark of Born's science. To be sure, he did other things that were important. For example, he and Oppenheimer produced an approximate scheme for treating molecules in quantum theory that we still teach and use. But the fundamental advances were made by the generation of Born's students and then their students. In 1930, he and Jordan published the first of what was intended to be a two-volume work on quantum mechanics. They made the mistake of devoting the first volume to matrix mechanics, which inevitably makes their discussion more mathematical and usually more difficult. Pauli wrote a scathing review which ended in typical Pauli fashion with the conclusion, "The layout of the book is excellent with regard to printing and paper."[38] The second volume, which would have discussed wave mechanics, was never written. Over the years, Born wrote several books. The two that I like the best, *Einstein's Theory of Relativity* and *Atomic Theory*, are still in print. I first learned the quantum theory from the latter. In later editions he tried to make it contemporary, but, by now, that part of the book must be very much out of date. He also wrote a notable memoir on optics.

The most satisfactory part of Greenspan's book — it involves the least science — is her account of Born's departure from Germany and his years abroad. Until I read Greenspan's book, I had not realized that as early as 1930, Göttingen, which as a former math major I always had romantic feelings about, was already a hot-bed of Nazism. A substantially larger percentage of the town had voted for the Nazi party than the national average. In April of 1933, Born was dismissed from the university. He decided to leave Germany with his wife, two daughters, and a son. Some of his Jewish colleagues decided to remain. They could not believe that the rise of Nazism was more than a temporary aberration. Germany, after all, was the country of Beethoven, Bach, and Goethe. It was also, alas, the country of Wagner. Einstein suggested that Born go to Palestine. Born wrote back that Palestine would be his last choice. Neither he nor his wife identified themselves as Jews. Cambridge University offered him a three year lectureship in mathematics which he accepted. Once this was decided, Born tried to arrange positions out of Germany for all his younger colleagues. With some he had success and with some he didn't. Most of the latter died in Germany.

In 1936, Born accepted a professorship at the University of Edinburgh where he remained until his retirement in 1953, the year before he won the Nobel Prize. While he was still in Cambridge, he collaborated with a young Polish physicist named Leopold Infeld who went on to work with Einstein in Princeton. By this time the quantum theorists at the frontier had gone beyond ordinary quantum mechanics and were applying the theory to electromagnetic fields — quantum electrodynamics. Quantum field theories had enjoyed some successes. For example, the scattering of electrons and photons could be calculated. But once one went beyond the first approximation, disasters appeared in the form of infinite results for quantities that should have been well-defined and finite. Real progress was not made until after the war, but Born and Infeld proposed a solution and wanted to make the electromagnetic equations nonlinear.[39] As far as I can tell, this modification did not contribute to the infinity question. But when I looked up this work on the Web, I was surprised to see that their equation is now a hot topic of study among string theorists.

Once he got to Edinburgh, Born once again attracted students. One of the most interesting was Klaus Fuchs. At this time Born had another idea, which as far as I know, went nowhere. He noticed that the canonical commutation relations have an obvious symmetry. If you take the position matrix and replace it by the momentum matrix and also replace the momentum matrix by the negative of the position matrix, the commutation relations remain unchanged. This symmetry also holds for the matrix that represents a particle's orbital angular momentum. Born persuaded himself that there was something profound about this symmetry and gave it a name, the principle of reciprocity. Although he and Fuchs worked on this problem, nothing much seems to have come of it. During the war Fuchs was reluctantly — the reluctance was his — made part of the British contingent that went to Los Alamos. By the time the war ended, he had given to the Russians the detailed plans for the construction of the plutonium bomb that was tested in Alamogordo in the summer of 1945. No one, certainly not Born, had any idea that he was a spy.

In addition to their disagreement about the quantum theory, Born and Einstein had profound differences of opinion about Germany. Einstein had almost an Old Testament view. He said that the Germans were a nation of mass murderers and he would have nothing to do with any of them. Born, some of whose immediate family had been exterminated, had a more nuanced view. We have already mentioned that he would not write an exonerating letter for Jordan. His feelings about Heisenberg are, at least to me, somewhat unclear. There is a persistent story that when Born went to visit Göttingen in the early 1950s, allegedly to try to get his old job back, Heisenberg spit on the ground. Heisenberg certainly said some incredibly foolish and insensitive things, but this sort of behavior does not ring true, and

Greenspan did not find any evidence for it. I would imagine, considering the background, that there may well have been some tension between them. In any case, Born was awarded the honor of "freedom of the city" by Göttingen in 1953, which he seemed to have had no trouble accepting. By this time he had decided to return to Germany to live. He and his wife settled in the spa town of Bad Pyrmont not far from Göttingen. Born gave as one of his principal reasons for returning to Germany, economics. He said that his pension from Edinburgh was small, and that he would be living on the pension that he now got from Göttingen. I suspect that this justification was only a part of the reason. He could have used this pension elsewhere, and he had the Nobel Prize money as well as income from his books. I think that he and his wife felt German. It was their mother tongue. Born had fought for his country in the First World War. Germany had been their home and would remain their home for the rest of their lives. Many of Born's friends and colleagues did not understand, but nostalgia for a lost homeland is a very powerful emotion. Born and his wife were buried in Göttingen. On their gravestone is carved

$$qp - pq = h/2\pi I$$

Notes

[1] Some of the more notable biographies include the following: Albrecht Fölsing, *Albert Einstein*, Penguin Books, New York, 1998; Abraham Pais, *Niels Bohr's Times*, Oxford University Press, New York, 1991; Walter J. Moore, *Schrödinger, Life and Thought*, Cambridge University Press, New York, 1992; Abraham Pais, Maurice Jacob, David I. Olive, and Michael F. Atiyah, *Paul Dirac: The Man and His Work*, Cambridge University Press, New York, 2005, David C. Cassidy, *Uncertainty; The Life and Science of Werner Heisenberg*, W. H. Freeman and Co., New York, 1992.

[2] The play in question is Michael Frayn, *Copenhagen*, Anchor Books, New York, 2001.

[3] Nancy Thorndike Greenspan, *The End of a Certain World*, Basic Books, New York, 2005.

[4] Reference 1, p. 144.

[5] Reference 1, p. 2.

[6] Barbara Goldsmith, *Obsessive Genius: The Inner World of Marie Curie*, Atlas Books, W. W. Norton & Company, New York, 2005.

[7] Goldsmith does not seem to understand that krypton was also produced in the fission whose mass Meitner had to know.

[8] Reference 4, p. 225.

[9] Reference 1, p. 62.

[10] For a wonderful view of Göttingen see John Derbyshire, *Prime Obsession*, Joseph Henry Press, Washington, DC, 2003. In contrast to the two books I have mentioned, Derbyshire's book, which concerns Riemann's hypothesis about the zeros of the Riemann function, is a model of how a popular science book should be written.

[11] This quote is from Constance Reid, Hilbert Springer, New York, 1996.

[12] *The Born-Einstein Letters: Friendship, Politics and Physics in Uncertain Times*, with an introduction by Kip Thorne and Diana Buchwald, Mc-Millan, New York, 2005.

[13] Reference 1, p. 61. The italics are mine.

[14] A. Einstein, "Strahlungs-Emission und — Absorption nach der Quantentheorie," Verh. Dtsch. Phys. Ges. 18, 318 (1916).

[15] A good account can be found in David Bohm, *Quantum Theory*, Dover, New York, 1989. Bohm compares the answer found in this way with the real quantum mechanical answer and shows that at least in simple cases they agree surprisingly well.

[16] A account of this chronology can be found in David Cassidy, *Uncertainty: The Life and Science of Werner Heisenberg*, W. H. Freeman, New York, 1992.

[17] I have been greatly helped by two papers that I would suggest reading in the following order. First, William A. Fedak and Jeffery J. Prentis, "Quantum jumps and classical harmonics," *Am. J. Phys.* 70(3), 332–344, 2002; and Ian J. R. Aitchison, David A. MacManus and Thomas M. Snyder, "Understanding Heisenberg's 'magical' paper of July 1925: A new look at the calculational details," *ibid.* 72(11), 1370–1379, 2004; A translation of the original paper can be found in *Sources of Quantum Mechanics*, edited by B. L. van der Waerden, Dover, New York, 1967, pp. 261–276.

[18] See van der Waerden, Ref. 13, p. 261.

[19] Here and in what follows I will ignore the following subtlety. In general these transition amplitudes are complex numbers and the squares are absolute values squared. For purpose of this essay I will treat them as if they were real numbers. This will not affect anything.

[20] Reference 1, p. 125.

[21] See van der Waerden, Ref. 13, p. 266.

[22] For a discussion see Marshall Bowen, and James Coster, "Born's discovery of the quantum-mechanical matrix calculus," *Am. J. Phys.* 48(6), 491–492, 1980.

[23] These are clearly spelled out in Aitchison *et al.*, Ref. 13, Appendix A.

[24] The time dependence of $x_{nm}(t)$ is $e^{i\omega}_{nm}t$. The velocity is the first time derivative hence the factor of i, while the acceleration is the second time derivative so you get i^2.

[25] See van der Waerden, Ref. 13, p. 40.

[26] Most of it is reproduced in translation in van der Waerden, Ref. 13. For reasons that are unclear van der Waerden leaves out the last few pages of the original which Jordan later complained about.

[27] Van der Waerden, Ref. 13, p. 277.

[28] I am grateful to Helmut Rechenberg for remarks on Jordan's career and other points in my review.

[29] This full text of this and many other wartime pronouncements of German scientists can be found in Physics and National Socialism, edited by Klaus Hentschel, Birkhäuser, Boston, 1996, p. 272.

[30] I am grateful to Peter Goddard for supplying this and for his comments.

[31] Later in 1926, Dirac published a second paper in which he used matrix methods to discuss the hydrogen atom. See P. A. M. Dirac, "Quantum mechanics and a preliminary investigation of the hydrogen atom," *Proc. R. Soc. London, Ser. A* 110, 561 (1926). In this paper he does refer to Born and Jordan.

[32] Reference 1, p. 191.

[33] M. Born, "Zur Quantenmechanik der Stoßvorgänge," *Z. Phys.* 37, 863– 867, 1926; "Quantenmechanik der Stoßvorgänge," *ibid.* 38, 803–827, 1926.

[34] One again I am ignoring the fact that these functions are in general complex numbers so that you must take the absolute value squared.

[35] This quotation can be found in Abraham Pais, *Subtle is the Lord*, Oxford University Press, New York, 1982, p. 442. The first bracket is mine and the other two are Pais's.

[36] Reference 30, p. 443.

[37] See M. Born, "The statistical interpretation of quantum mechanics," in *Nobel Lectures in Physics 1942–1962* Elsevier, Amsterdam, 1964.

[38] Reference 1, p. 159.

[39] M. Born and L. Infeld, "Foundations of the new field theory," *Proc. R. Soc. London, Ser. A* 144, 425–451, 1934.

8. The Quantum Measurement Problem

Rhyme-royal's difficult enough to play.
But if no classics as in Chaucer's day,
At least my modern pieces shall be cheery
Like English bishops on the Quantum Theory.

W.H. Auden,
Letter to Lord Byron, 1934

This essay presents a kind of "tasting menu" for the various approaches to the quantum measurement problem. This is not often taught in standard quantum mechanics courses, so the student is not exposed to this fascinating and important subject. Perhaps this somewhat informal presentation will encourage further study.

1. The Problem

In his 1951 text *Quantum Theory*[a] David Bohm presented a treatment of quantum mechanical measurement that anticipates to some extent the now-fashionable notion of "decoherence." Bohm considered this in the context of a Stern–Gerlach experiment in which the

---

[a]David Bohm, *Quantum Theory*, Prentice-Hall, New York, 1951. When I was learning quantum mechanics this was the only text that dealt with these questions. We were advised to skip these parts of the book during the course.

spin of a particle is measured. The initial wave function can be written as

$$\Psi(z,0) = f_o(z)(c_+ u_+ + c_- u_-),\tag{1}$$

where the u's are the spinors and the c's are their complex coefficients. Over the course of time, in the presence of an inhomogeneous magnetic field in the z-direction, this function evolves continuously, using the Schrödinger equation, into

$$\Psi(z,t) = f_+(z,t)u_+ + f_-(z,t)u_-.\tag{2}$$

We imagine that the magnetic field H is expanded in the z-direction so that

$$\mathrm{H} \sim \mathrm{H}_0 + z\mathrm{H}_0'.\tag{3}$$

Thus the evolved solutions can be written as

$$f_+(z,t) = u_+ f(z,0)\exp-(i\mu(H_0 + zH_0')t/\mathrm{h}),\tag{4a}$$

and

$$f_-(z,t) = u_- f(z,0)\exp(+i\mu(H_0 + zH_0')t/\mathrm{h}).\tag{4b}$$

Here $\mu = -e\hbar/2\,mc$ is the magnetic moment and the sign difference between (4a) and (4b) reflects the interaction term which is proportional to $\sigma \cdot H$ where σ is the Pauli spin matrix.

These two equations, I am going to argue, are as far as one can go using only Schrödinger quantum mechanics. But they do not yet constitute a "measurement." To proceed to a measurement we must go outside this framework. What Bohm did is to introduce semiclassical approximations. If one chooses to say so (Bohm certainly didn't) these are a form of disguised "hidden variables." Classical properties are attributed to the particles. Here is a sketch of how it works.

A particle spends a small time Δt in the neighborhood of one of the magnets. It is acted on then by a force for say positive spin

$$\Delta\mathrm{p}/\Delta\mathrm{t} = -\mu\mathrm{H}_0'.\tag{5}$$

For negative spin we have the opposite sign. Hence the apparatus acts as a momentum filter and if we were to measure the momenta after the particles' encounter with the magnets we would have measured their spin. But, we are in effect treating the particles as classical objects as far as the measurement process is concerned.

The Stern–Gerlach experiment, certainly as performed by Otto Stern and Walter Gerlach, did not involve the separation of momenta but rather the separation of orbits in space. To deal with this, Bohm constructed wave packets for the two spin states. He again resorted to a semi-classical approximation to describe how these packets move. He argued that they separate spatially depending on the spin and they are located far enough from each other so that there is essentially no spatial overlap. This implies that, if we compute the expectation value of any function of the spin averaged over space, the cross terms involving the products of the packets can be neglected. The probabilities become classical.

If we focus on the momentum filter, then, for it to work, the momentum imparted by the magnetic force must be greater than the uncertainty principle momentum Δp; i.e., $\mu H_0' \Delta t > \Delta p$ or $\mu H_0' \Delta t / \Delta p > 1$. We have seen that the phase in the wave functions is proportional to $\mu H_0' \Delta t \Delta z / \hbar > \mu H_0' \Delta t / \Delta p > 1$. So under the circumstances where a measurement is possible, the phase becomes very large. Let us consider this in terms of the density matrix that characterizes the mixture. This is a 2×2 matrix whose diagonal terms are of the form $|\Psi_+\rangle\langle\Psi_+|$, and whose off-diagonal terms are of the form $|\Psi_-\rangle\langle\Psi_+|$, where the $+$ and $-$ refer to the up/down spins. In the diagonal terms the phases cancel while in the off-diagonal terms they add. This is important because when the particles enter the magnetic field the phases become very large so the cross-terms oscillate to zero. Thus the density matrix is effectively diagonalized. This however, still does not constitute a measurement.

A measurement occurs when some process causes one of the diagonal elements to become zero. This is often referred to as the "collapse of the wave function." I will not discuss here the difficult problem of finding a mechanism that does this. I am only going to point out that whatever mechanism this is, it cannot obey the Schrödinger equation linking the system to the apparatus. To make clear why this is so let me call the projection operator that collapses the wave function P. P is Hermitian and idempotent; i.e., $P^2 = P$. Let us look at the norm of the projected wave function

$$\langle P\Psi | P\Psi \rangle = \langle \Psi | P^2 | \Psi \rangle = \langle \Psi | P | \Psi \rangle \neq \langle \Psi | \Psi \rangle. \qquad (6)$$

This means that the collapse cannot be described by a Schrödinger equation. The solution to this equation is symbolically $\exp(-iHt/\hbar)$ $\psi(0)$ where H is a Hermitian Hamiltonian. The operator $\exp(-i Ht/\hbar)$ is unitary and does not change the norms of vectors. This is the quantum mechanical measurement problem, or at least one part of it.

There is the very important matter of "Born's rule." This was presented by Born in 1925. It is the proposition that if the wave function has been projected onto one of the eigenstates of the Hermitian operator representing some observable, then the absolute square of the scalar product of this projection and the wave function represents the probability of finding the eigenvalue corresponding to that eigenstate in the measurement. In most of the interpretations that I am going to discuss, this is an additional postulate. Whether in any of them it can be proven as a theorem is an open question which I will not discuss.[b] I will now consider some proposed solutions to the collapsed wave function problem.

2. Solutions

2.1. *The George Dyson solution*

This solution would, I think, garner the largest number of votes among working physicists. It is based on a conversation that George's father Freeman reported to me a number of years ago when George was a young child. His sister Esther, who is a year or so older, was interested in establishing her intellectual superiority like any older sister. One day, she announced to George that she now understood how a boat rows. "You take the rowers," she said, "and make holes in the water into which the boat falls." To this George replied, "I do not understand how a boat rows, but I can row it anyway." Substitute "quantum theory" for "boat" and you have the solution.

---

[b]I am grateful to Steve Adler for discussions on this.

2.2. *Bohm's 1951 textbook*

In his textbook Bohm presents the phase discussion I just went through in the introduction. Since, as far as the text is concerned, he does not represent this as a problem, he offers no solution. He does remark that measurement is an irreversible process in the thermodynamic sense. He even compares it to cell growth. But he does not discuss the fact that measurements cannot be described by the Schrödinger equation nor does he really explain what a measurement is in general, as opposed to some other form of interaction, one that can be described by a Schrödinger equation. We might call this the Alfred E. Neuman or "What, me worry?" solution.

2.3. *Bohr*

Bohr was adamant that measurement is a classical process.[c] We must be able to describe a measurement in everyday language. It is the only language we have. But in quantum mechanics we must use this language carefully. We can present a classical description of a momentum experiment and a classical description of a position experiment and we can present a classical description of an experiment that would measure both of them with arbitrary precision. But such an experiment would violate the laws of quantum mechanics. Position and momentum are two "complementary" attributes. One of my favorite Bohr sayings is that one should not speak more clearly than one thinks. One should also not speak more clearly than quantum theory allows. But the problem for many students of Bohr was, where does quantum world leave off and the classical world begin? Bohr does not seem to have written anything very clear about this except to imply that quantum mechanics deals with small things while classical physics deals with big things. The "cut" that separates the classical and quantum worlds is left ambiguous. The same issue arises for people who suspect that quantum theory applies only to patches of the universe. What is a patch? Where are they located? Such people

---

[c]An especially nice discussion of these matters can be found in Andrew Whitaker, *Einstein, Bohr and the Quantum Dilemma*, Cambridge University Press, New York, 1996.

abandon the hope of finding a wave function that applies to the universe at large and its history.

2.4. *Bohm 1952*

Soon after it was published, Einstein read a copy of Bohm's book. He called Bohm, who was also in Princeton, and said that he would like to talk with him about it. I do not know exactly what Einstein said to Bohm about his objections to the theory, but Bohm decided that the old man had a point and proceeded to revive an old interpretation of quantum theory that was put forth by de Broglie in the late 1920's. Pauli seems to have scared de Broglie out of it with some acerbic criticism that involved the treatment of inelastic scattering. Bohm, on the other hand, had three decades of the development of quantum theory behind him so he was not scared easily. The idea is that you have a classical position variable which I will call Q. (It is easy to generalize this to several such variables.) This is the position of a particle that moves along a classical trajectory that is determined by the following pair of equations which I will write in one spatial dimension. There is the "guide wave" $\psi(x, t)$ which obeys the Schrödinger equation

$$\partial/\partial t \Psi(x, t) = -i/\mathrm{h} H \Psi(x, t). \qquad (7)$$

The solution is fed into the equation for the Q's

$$\partial/\partial t Q(x, t) = (\mathrm{Im}\Psi^*(Q(x, t)\partial/\partial x Q(x, t))/\Psi^*(Q(x, t)\Psi(Q(x, t)). \qquad (8)$$

Each trajectory is classical and deterministic but their relative probability is given by $|\psi|^2$. This is how the quantum mechanical uncertainties are reproduced. This probability interpretation is an assumption which, it is probably fair to say, would not have been thought up if quantum mechanics had not made the same assumption. There is no measurement problem here. Measurements like the Stern–Gerlach experiment are not different from any other kind of interaction. Just follow the trajectories. But there is another problem that John Bell was the first to point out.[d] If you have more than one free particle then the trajectories are factorizable. Each particle follows its own path. But the wave function in the presence of interactions is not. This introduces an essential non-locality into the theory. If a butterfly flaps its wings somewhere, in principle this influences the wave function instantaneously. It is also not clear how to make the theory relativistic. We may have replaced a mystery by an enigma.

2.5. *Changing the Schrödinger equation*

I once heard a lecture by Wigner on quantum measurements. This was about the time in the early 1960's when he began publishing papers on this subject.[e] Wigner explained why he thought that "orthodox" quantum theory, as he called it, could not account for quantum measurements. The essence is that the Hamiltonian H is a linear operator and so is the operator $\exp(-iHt/\hbar)$ (to see this expand the exponential). Thus if you began with a wave function that was a product of the function that represented the apparatus and a linear combination that represented the system, you would end up after this time evolution with a linear product. There would be no collapse of the wave function. Wigner thought that the Schrödinger equation would have to be modified so that such a collapse could be accounted for. He also felt that human consciousness might play a crucial role. An experiment is not an experiment, he felt, unless it was registered by an observer. The role of consciousness in quantum mechanics has been much discussed. For the purposes of this presentation I will adopt a Newtonian attitude of *consciousness non fingo*. Much more recently Ghirardi, Rimini and Weber[f] proposed such an addition to the Hamiltonian, an interaction that would periodically and randomly collapse the wave function. This introduces a new constant of nature λ that is the rate of collapse. (In their model there is a second model-dependant constant which I will not

---

[d]See for example p. 1 *et. seq.* in J.S. Bell, *Speakable and Unspeakable in Quantum Mechanics*, Cambridge University Press, New York, 2004.

[e]See for example E.P. Wigner, *Am. J. Phys.* **31**, 6 (1963).

[f]G.C. Ghirardi, A. Rimini, T. Weber, *Phys. Rev. D* **34** p. 470.

discuss.) What one needs is for the effective rate to be slow when microscopic processes are involved and fast when macroscopic processes are at play. In a spin measurement, for example, you begin with an entangled state of two quantum mechanical particles, but after the measurement you have single quasi-classical branches. The way this happens in the model is to take $\lambda \sim 10^{-16}\,\mathrm{sec}^{-1}$ and then to note that when macroscopic processes are involved this will be boosted by something like Avogadro's number so that the rate becomes something like $10^7\,\mathrm{sec}^{-1}$. In this scheme energy is not quite conserved but they argue that the violations are so small as to be unobservable. I am not aware of any other introduction in physics of a fundamental constant in response to a philosophical imperative. One wonders if they have replaced an enigma by a mystery.

2.6. *Worlds, paths and histories*

Let me begin by discussing the past. With or without quantum mechanics it differs from the future. The past is about what happened and the future is about what might happen. Intuitively we believe that retrodiction is unique and deterministic. The question is if quantum mechanics can account for this. To see the problem, let us consider once again the Stern–Gerlach experiment. Once one of the components of the wave function has been projected out, its evolution in time will once again follow a Schrödinger equation. But now suppose we run this scenario backwards in time. A projection operator does not have an inverse, so we cannot reconstitute the mixed state. The theory does not determine what had been the case prior to the measurement. There is more than one possible past. In his book *The Physical Principles of the Quantum Theory*[g] Heisenberg discusses the past in somewhat different terms. He writes,

"This formulation [of the quantum theory] makes it clear that the uncertainty relation does not refer to the past; if the velocity of the electron is at first known, and the position exactly measured, the positions for times previous to the measurement may be calculated.

---

[g]Werner Heisenberg, Dover Press, New York, 1949. I am grateful to Eugen Merzbacher for pointing out this reference. See p. 20.

Then for these past times $\Delta p \Delta q$ is smaller than the usual limiting value, but this knowledge of the past is of a purely speculative character, since it can never (because of the unknown change in momentum caused by the position measurement) be used as an initial condition in any calculation of the future progress of the electron and thus cannot be subjected to experimental verification. It is a matter of personal belief whether such a calculation concerning the past history of the electron can be ascribed any physical reality or not."

Yes, but what about **our** past histories? Can these be accounted for by the quantum theory? If so, how? This is a theme that arises in the interpretation of the theory I am now going to discuss.[h]

If I were writing a history of this interpretation I would start with the paper by Dirac which he wrote in 1932 and published in the Russian journal *Physikalische Zeitschrift der Sowjetunion* in 1933,[i] At the time he was thinking about the paper, which is called "The Lagrangian in Quantum Mechanics," he was on a trip to Russia with other physicists. One of his fellow travelers was my teacher Philipp Frank. Professor Frank often spoke about a boat trip on the Volga they all took. He showed me some pictures. It seems as if Dirac said on that occasion that he had now understood the role that the Lagrangian played in quantum mechanics. The role of the Hamiltonian was well known, but where was the Lagrangian? Dirac's key observation is contained in Equation (9) of his short paper which I present below.

He considers the time evolution of the scalar product of a state characterized by a coordinate q at time t and the same coordinate at a later time T. He writes

$$\langle q_t | q_T \rangle \quad \text{corresponds to} \quad \exp\left[i \int_T^t Ldt/\hbar \right]. \tag{9}$$

---

[h]I am very grateful to Jim Hartle for comments on this. I strongly recommend his paper *Quantum Pasts and the Utility of History*, arXiv:gr-q/9712001.

[i]The paper is reproduced in *Feynman's Thesis* edited by Laurie Brown, World Scientific, Singapore, 2005. The essential arguments are repeated in Dirac's *The Principles of Quantum Mechanics*, Third Edition, Oxford University Press, Oxford, 1947, p. 125 *et. seq.*

Here L is the Lagrangian. Many years later Feynman asked Dirac if he had worked out what "corresponds to" meant; i.e. what the constants were. He had not, but Feynman did in the PhD thesis he wrote with Wheeler. One thing Dirac did do was to go to the classical limit in which Planck's constant tends to zero. The exponent oscillates wildly unless the integral, which can be broken up as integrals over intermediate times (or "paths")

$$\langle q_t | q_T \rangle = \int \langle q_t | q_m \rangle \, dq_m \ldots \langle q_2 | q_1 \rangle \, dq_1 \langle q_1 | q_t \rangle \qquad (10)$$

is stationary along these paths.

But this is just the condition, as Dirac showed, that gives the classical equations of motion. In his thesis, Feynman works these things out in detail and gives some specific examples such as the anharmonic oscillator. Wheeler was so smitten by this that he decided that if he explained it to Einstein, Einstein would be converted. Of course, he wasn't. Einstein's problems with quantum theory had nothing to do with formalism.

I do not think that Feynman attached any special reality to these alternate paths. For this we can thank Hugh Everett, another student of Wheeler's. In 1957, he published his thesis in which each of these paths corresponds to another "world." A measurement is just a fork in the road and, as they say, if you come to a fork in the road, take it — take all of them. There is no collapse of the wave function. A motivation for Everett was to have an interpretation of quantum theory that would make it applicable to the universe at large. There are no patches. A veritable cottage industry has grown up from Everett's thesis. I am most familiar with the work of Murray Gell-Mann and James Hartle[j] who give an account of how these forks are actually chosen and why some of them are classical or nearly so. A crucial point is that the alternate quasi-classical histories decohere. This is in general a different form of decoherence than the wave function decoherence we have discussed above. Here paths decohere. This happens because the system collides with various backgrounds.

---

[j] Again a very clear popular account can be found in Murray Gell-Mann, *The Quark and the Jaguar*, Owl Books, New York, 2002.

The Moon follows a quasi-classical path, and is there when we don't observe it because it has collided with microwave photons from the Big Bang, for example. It follows a decoherent quasi-classical path. One thing about this program that recommends itself to me is that it attempts to account for the past. Like the usual quantum theory it allows for the possibility of several pasts. Ours happens to be one of them and, fortunately, many of us share it, or so we claim.

2.7. A mathematician's solution

In 1932 John von Neumann published his *Mathematische Grundlagen der Quantenmechanik.*[k] von Neumann attempted to present the theory as a series of deductions from a set of axioms. Needless to say, the axioms and definitions were taken from physics. He distinguished between two kinds of operations in the theory. Type 1 deals with measurements. We suppose that there is a set of eigenstates Φ_n of some Hermitian operator that represents an observable. We call P_n the projection onto the nth-eigenstate. Then von Neumann defines the measurement operation relating to some operator U by the relation

$$U \otimes U' = \sum_{n=1}^{n=\infty} (U\Phi_n, \Phi_n) P_{\Phi_n}. \tag{11}$$

This is the collapse of the wave function.

It is distinguished from a Type-2 operation which takes the form

$$U \otimes U_t = \exp(-i/\hbar Ht) U \exp(i/\hbar Ht), \tag{12}$$

where as usual H is the Hamiltonian. von Neumann then goes on to ask why there needs to be two sorts of operations. He writes,

"First of all, it is noteworthy that the time dependence of H is included in (12) (in the manner described there), so that one would expect that (12) would suffice to describe the intervention caused by a measurement: Indeed, a physical intervention can be nothing else than the temporary insertion of a certain energy coupling into the observed system, i.e.,

---

[k]The English translation *Mathematical Foundations of Quantum Mechanics* was published by Princeton University Press, Princeton, New Jersey, 1955.

the introduction of an appropriate time dependency of H (prescribed by the observer). Why then do we need the special process (11) for the measurement? The reason is this: In the measurement we cannot observe the system S by itself, but must rather investigate the system S + M, in order to obtain (numerically) its interaction with the measuring apparatus M. The theory of the measurement is a statement concerning S + M, and should describe how the state S is related to certain properties of the state of M (namely, the positions of a certain pointer, since the observer reads these. Moreover, it is rather arbitrary whether or not one includes the observer in M, and replaces the relation between the S state and the pointer positions in M by the relations of this state and the chemical changes in the observer's eye or even in his brain (i.e., to that which he has "seen" or "perceived").... In any case, therefore, the application of (12) is of importance only for S + M. Of course, we must show that this gives the same result for S as the direct application of (11) on S. If this is successful, then we have achieved a unified way of looking at the physical world on a quantum mechanical basis ...."[1]

von Neumann, like Wigner, thought that consciousness played an important role.

At the end of his book Von Neumann introduces a toy model of a measurement. He considers a particle of momentum p whose position x we want to measure. This he couples to an apparatus of momentum P with a coupling of the form g(t) × P. He solves the time dependant Schrödinger equation in the approximation in which the kinetic terms in the Hamiltonian are neglected — the "impulse" approximation. It is I think very instructive to do the same problem in the same approximation in the Heisenberg representation. What we want to do is to see how a change in position of the apparatus — the position of a "pointer" — determines the position of the object under observation. Let us call the position of the pointer R(t) and let us assume that the impulse is applied during the interval 0–T. Thus, using the canonical commutation relations and ignoring the kinetic term,

$$R(T) - R(0) = \int_0^T \frac{dR}{dt} dt = \int_0^T i/\mathrm{h}[H, R] dt = gx. \qquad (13)$$

---

[1] von Neumann, op. cit. 351–352.

Here

$$g = \int_0^T g(t)\,dt.$$

Thus we have related the change in the pointer position to the position of the particle. The reader may well ask, where in this are the interactions of type 1 that von Neumann was so careful to describe? It looks as if we have performed a measurement using only orthodox quantum mechanics. However, I claim that this is not a measurement. It is more like a preparation for a measurement. To make a measurement we must register the result using for example a photographic plate. It is this registration that is not reversible and is not described by von Neumann's model. He comments,

"The further question, whether the

$$\Delta = \exp(i/\mathrm{ht}H)$$

corresponding to simple and plausible measuring arrangements also have this property, shall not concern us."[m]

I was somehow led to recall Hilbert's remark that theoretical physics was much too difficult for physicists. It may also be too difficult for mathematicians.

This then is my admittedly subjective brief overview of the quantum measurement question. There are more theories than the ones I have discussed and much more to be discussed about each of the ones I have included. You can think of this as a survey of terrain from the air. If you really want to survey the terrain you have to land the plane, get out and touch the ground.

---

[m]von Neumann, op. cit. p. 441.

Part II. Nuclear Weapons

9. Oppenheimer

"That is what novels are about. There is a dramatic moment and the history of the man, what made him act, what he did, and what sort of a person he was. That is what you are really doing here. You are writing a man's life."

I. I. Rabi[1]

"The story I told Pash was not a true story. There were not three or more people involved. There was one person involved. That was me. I was at Los Alamos. There was no one else at Los Alamos involved. There was no one at Berkeley involved. When I heard the microfilm or what the hell, it didn't sound to me as to this reporting anything that Chevalier had said, or at that time the unknown professor had said. I am certain that was not mentioned. I testified that the Soviet consulate had not been mentioned by Chevalier.

That is the very best of my recollection. It is conceivable that I knew of Eltenton's connection with the consulate, but I believe I can do no more than say the story told in circumstantial detail, and which was elicited from me in greater and greater detail during this [hearing] was a false story. It is not easy to say that.

Now, when you ask for a more persuasive argument as to why I did this than that I was an idiot, I am going to have more trouble being understandable."

J. Robert Oppenheimer[2]

In his book *Oppenheimer: The Story of a Friendship*,[3] Haakon Chevalier writes that he is not sure of the exact occasion when he and

Oppenheimer first met. He is sure of the year: 1937. Oppenheimer was thirty-three and already acknowledged to be one of the best theoretical physicists of his generation. He held joint appointments at Berkeley and Caltech and had a comfortable private income from his family. Chevalier, who was born in New Jersey of mixed French and Norwegian parents, was teaching French literature at Berkeley. He had written an important book on Anatole France and had translated André Malraux. Later he would write novels. His description of the Oppenheimer of 1937 is written with a novelist's eye. In it I recognize the Oppenheimer I got to know twenty years later. Chevalier writes:

> "I can remember my earliest impressions of him at this stage. He was tall, nervous and intent, and he moved with an odd gait, a kind of jog, with a great deal of swinging with his limbs, his head always a little to one side, one shoulder higher than the other. But it was the head that was most striking: the halo of wispy black curly hair, the fine sharp nose, and especially the eyes, surprisingly blue, having a strange depth and intensity, and yet expressive of candor, that was altogether disarming. He looked like a young Einstein, and at the same time like an overgrown choir-boy. There was something both subtly wise and terribly innocent about his face. It was an extraordinarily sensitive face, which seemed capable of registering and conveying every shade of emotion. I associated it with the faces of apostles, either imagined or remembered from Renaissance paintings. A kind of light shone from it, which illuminated the scene around him."[4]

Of course, twenty years later his hair had grown white and was cut very short. He sometimes had a look of martyrdom, and he was capable of being quite cruel when he thought his time was being wasted. He was no longer really doing physics, but what struck me was how important other physicists — many who had been much more creative than he — felt his approval was.

The thing that brought Oppenheimer and Chevalier together initially was their common interest in left-wing causes. In reading Chevalier himself, as well as what has been written about him, I am not absolutely clear as to whether he was, or was not, a member of the Communist Party; most likely he was. In any event, he was a fellow traveler, *pure et dure*. Oppenheimer, on the other hand, until some time shortly before he met Chevalier, had had an almost total indifference to politics. He reported that until 1936 he had never voted in a presidential election. He was not aware of the Depression until some of his gifted students could not get jobs. Even though he had earned his Ph.D. in Germany in the late 1920s, he was more or less indifferent to Hitler until his own relatives had to flee Germany. He claimed that he never read a newspaper or a news magazine, and did not even have a radio or a telephone. Yet, from 1936 until Pearl Harbor, he suddenly had an intense period of political activity, largely in the service of causes such as the Loyalists in the Spanish Civil War and the organization of migratory workers, that were closely linked to Communist interests. What brought about the change? At his hearing before the Personnel Security Board of the Atomic Energy Commission, in 1954, Oppenheimer suggested an answer. He recalled that in the spring of 1936, some friends introduced him to a woman named Jean Tatlock, who was the daughter of an English Professor at Berkeley. She was studying psychiatry. They began a stormy affair that lasted until he married Katherine "Kitty" Puening in November of 1940. Tatlock was a member of the Party, and very committed to the activities that Oppenheimer soon became involved with. Through her he met others who were either Communists or fellow travelers. These included Kitty, whose husband Joe Dallet had died fighting for the Loyalists as part of the Abraham Lincoln Brigade. She had been a member of the Party during her marriage. About this time, and against Oppenheimer's advice, his brother Frank and his wife joined, as had several of Oppenheimer's students. This raises the obvious question of whether Oppenheimer himself, despite unambiguous and consistent denials, had been a member. That he had is the contention of the book Brotherhood of the Bomb, by the historian Gregg Herken. I will come back to explain why I find neither this, nor much of Herken's portrayal of Oppenheimer, persuasive, but before I do so I want to continue with Chevalier, as this relationship led, in Rabi's words, to the "dramatic moment in the history of the man."

From Chevalier's account, his friendship with Oppenheimer, and later with Oppenheimer's wife, became very close. Oppenheimer had a deep interest in literature. He once told me that he had read all of Proust's novel while taking a long bicycle tour around Corsica, an anecdote that Chevalier also[11] relates in his book. Tatlock introduced him to the love poetry of Donne. Herken claims that it was with thoughts of Tatlock that he called the first atomic bomb test, which took place in New Mexico on July 16, 1946, Trinity — this from Donne's fourteenth "Holy Sonnet:" "Batter my heart, three person'd God ...." He writes that it was a "secret tribute to Jean Tatlock, who had committed suicide at her San Francisco apartment the previous January." He is off by a year. After taking sleeping tablets, she drowned herself in the bathtub in January of 1944. That aside, the evidence he cites does not stand up. When, in 1962, General Groves asked him this question, Oppenheimer said that he was thinking of Donne's poetry. He wrote, "I did suggest it. Why I chose the name is not clear, but I know what thoughts were in my mind. There is a poem of John Donne, written just before his death which I know and love. From it a quotation:

'As West and East In all flat Maps — and I am one — are one, So death doth touch the Resurrection.'

That still does not make Trinity; but in another better known devotional poem Donne opens, 'Batter my heart, three person'd God.' Beyond this, I have no clues whatsoever."[5]

Herken gives as a reference a memoir by the poet Edith A. Jenkins, entitled *Against a Field Sinister*. Here is what Jenkins wrote: "Jean introduced Oppenheimer to the poetry of John Donne. When the trial bomb went off in the Jornado del Muerto in New Mexico, Oppie dubbed it 'Trinity' after Donne's sonnet 'Batter my heart, three person'd God ...'.

That was approximately six months after Jean's death."[6] I was able to contact Ms. Jenkins by e-mail, and pointed out the misprint about the date her friend died. She confirmed to me that she knows of no evidence that the naming of "Trinity" had anything to do with Tatlock. Incidentally, one of the maddening features of Herken's book is the index. There is no entry under Donne and none under Auden, who also comes into the narrative as one of Oppenheimer's favorite poets. Later I will describe how I became witness to the only encounter between Oppenheimer and Auden. Chevalier once tried to introduce Oppenheimer to Malraux, who was visiting California. But it was only subsequent to Oppenheimer's hearing that Chevalier was able to arrange such a meeting, after which Oppenheimer made one of his typically delphic remarks, "Malraux has some understanding as to what science isn't. But he has no conception of what science is."[7] Malraux later commented that what Oppenheimer should have done at his hearing was to stand up and say "Je suis la bombe atomique!"[8] and then, presumably, to have walked out.

In late 1942, an event occurred that ultimately ruined both Chevalier's and Oppenheimer's lives. Herken supplies a clear chronology, although exactly who said what to whom is still murky. A British-born petroleum engineer named George Eltenton, who had lived in the Soviet Union and was a member of the Party, was approached by the third secretary of the Soviet consulate in San Francisco — a man named Pyotr Ivanov — to try to learn which of the physicists working at Berkeley on matters related to nuclear energy might be willing to share information with the Soviet Union. Ivanov knew that Eltenton had a relationship with Chevalier and that he, in turn, knew Oppenheimer. It is at this point that motive and intent become unclear. In his book Chevalier gives his version. He says he received a phone call from Eltenton, who wanted to see him on an important matter. When he went to Eltenton's house he learned that this was to sound him out about recruiting Oppenheimer. Chevalier says that Eltenton seemed nervous and unsure of himself and that he, Chevalier, told him in no uncertain terms that Oppenheimer would never agree to such a thing. In later depositions Eltenton, for whatever it is worth, confirmed this version of the story. The matter might have ended there except for what Chevalier did next. Soon afterwards, he and his wife were invited for dinner at the Oppenheimers'. In a private moment Chevalier told Oppenheimer about Eltenton, with the intent, he says, of simply warning him about possible espionage. He notes that, as far as he was concerned, this ended the matter. He writes, "We went back into the living room with the cocktail shaker, the gin and the vermouth, and joined our wives. I dismissed the whole thing from my mind."[9]

In the meanwhile a military man named Boris Pash had become involved with investigating security for the army. There was a turf-war between the army and the FBI, which was engaged in wire-tapping everyone who was suspect, including, eventually, Oppenheimer. By the spring of 1943, Oppenheimer had moved to Los Alamos as its director, but in August he came back to Berkeley for a visit. He casually mentioned to an army security officer that he had been contacted some months earlier about possible espionage. Promptly, he found himself confronted by a formal inquiry conducted by Pash, who had the colloquy recorded. It was at this point that Oppenheimer made a dreadful mistake. Instead of telling Pash the simple truth — whatever that was! — he invented a story, in order, he later said, to try to protect Chevalier, while at the same time conveying to the authorities his concern about espionage. He said that he had been approached by "intermediaries" — plural! — whom he would not name, to divulge information about the nuclear program, and that he knew for a fact that three other people had been approached. This appears to have been a total fabrication. Upon further questioning he mentioned Eltenton as one of the intermediaries. He hinted that a member of the Berkeley faculty had been involved, but refused to give the name. Pash reported the matter to General Groves, and he and his security officer, John Lansdale, interrogated Oppenheimer soon afterwards on a train trip. Oppenheimer still did not reveal the name of Chevalier, and Groves did not order him to do so. This changed the following December when Groves did insist on having the name of Oppenheimer's contact. He gave Chevalier's name, adding that he was "quite a Red." But instead of giving Groves what was probably the true version of the story — namely, Chevalier's — Oppenheimer introduced yet another fabrication. This time he admitted that there were not three people involved, but only his brother Frank. Now he had put two people in jeopardy — Chevalier and his own brother. Chevalier was pursued intensively, finally losing the opportunity to work in this country or even to hold an American passport. He took advantage of his dual citizenship, and moved to France. Frank Oppenheimer was also harrassed out of academic life and, for a while, took up ranching.

This, in essence, is the "Chevalier incident," but the question is why: Why did Oppenheimer behave in this way? It is at this point that I found Herken's often very instructive book a real disappointment. He does not seem to have any insight into Oppenheimer's character and, worse, makes assertions on the basis of what seems to me to be marginal evidence or indeed misinformation. I will come back to examples shortly, but first let me say what Herken tries to do. He presents a history of the development of the bomb in terms of three principals: Ernest Lawrence, Edward Teller, and Oppenheimer. Lawrence and Teller are pretty difficult to get wrong. Lawrence, the inventor of the cyclotron, was a scientific entrepreneur of the type that would be at home running a large corporation. Indeed, the science he did was much like a business enterprise. He simply wanted to build larger and larger machines, and in doing so missed many of the scientific discoveries made by others using much more modest equipment. Lawrence, who was a few years older than Oppenheimer, was born in South Dakota. His values were pretty straightforward, and while he admired Oppenheimer immensely for his scientific ability, he had little patience for his political and cultural activities. Teller, at least in my view, is a Dr. Strangelovian cartoon. I think the reader of Herken's book will come away with the feeling that he was even more devious and obsessed than one had imagined. Despite what he later claimed, it seems clear that he entered into his role in the Oppenheimer hearing with relish. But it is Oppenheimer who has the elusivity of a chameleon. I doubt that anyone who did not know him could get him right, and I am not sure that even people who knew him well actually understood him. An exception, as I will explain, was I.I. Rabi, who, as far as I am concerned, got him just right.

Although there is much I admire in his book, in my view Herken sometimes goes beyond the evidence, and occasionally distorts it, as we have seen in the case of Tatlock. I will give some additional examples. The first is contained in the caption to a picture of the Caltech physicist Richard Tolman and his wife Ruth. It reads "Richard and Ruth Tolman, 1941. Oppenheimer reportedly first earned Lawrence's disapproval when he seduced the wife of Professor Tolman at Caltech." Since this had not been mentioned in the text, I looked in the index. There is no entry under Ruth Tolman, but there is a gossipy footnote about Lawrence's learning at a cocktail party that Oppenheimer had had an affair with Ruth Tolman. Lawrence passed this on to Lewis Strauss, who added it to his catalogue of reasons for hating Oppenheimer. He noted this conversation in a memorandum to himself, remarking in words the same as the picture caption earlier identified. According to Dr. Lawrence, it was a notorious affair which lasted for enough time for it to become apparent to Dr. Tolman who died of a broken heart.

Clearly there are two things to be distinguished here. First, did Lawrence believe this, and second, is the story true? I have no reason to doubt that Lawrence believed it, and I have no doubt that this added to the reasons why he, Strauss, and Teller who was also informed wanted to destroy Oppenheimer. But I have several reasons for believing that it was not true — at least, not in the way that Lawrence described it. My complaint about Herken's book is that the reader is given no reason to question the validity of Lawrence's story as presented. Oppenheimer is branded, among other things, as a sordid adulterer, who ruined a colleague's life. When I read this contention, which was entirely new to me, I decided to check it out. This is not easy. All the principals are dead. Tolman died in 1948, and his wife Ruth in 1957, Lawrence in 1958, and Oppenheimer in 1967. There are not many witnesses left from the pre-war years when this affair supposedly took place. But there were a few. Among them was the physicist Robert Christy, who became Oppenheimer's student in 1937. When I asked Christy, he said that he had never heard of this, although it was well-known on the campus that Robert and Kitty had had an affair while Kitty was still married to Richard Stewart Harrison — a British doctor who became her second husband in 1938, after Joe Dallet died. This was confirmed by Tom Tombrello, the then chairman of the physics, mathematics and astronomy divisions at Caltech. He arrived there in 1961, when the premarital relationship between Oppenheimer and Kitty was still being discussed. He informed me that he had actually worked with Harrison, who subsequently remarried and went on to lead an entirely happy life. Tombrello also discussed the matter with Marge Lauritsen Leighton, the widow of Tom Lauritsen, a Caltech nuclear physicist. She said that when she was first married to Lauritsen they lived in the Tolmans' house with the Tolmans, and that she was a close friend of both the Tolmans and the Oppenheimers. She had never heard of any affair between Ruth Tolman and Robert Oppenheimer, and doubted if it had ever occurred. Moreover, Christy told me that when

Oppenheimer came to Caltech to teach, before the war, he always stayed in the Tolmans' house, which was on campus, and continued to do so even after he and Kitty were married. The least one can say is that this affair, if it ever existed, was certainly not "notorious." Everyone is in agreement that there is no evidence that Tolman died of a "broken heart." It was, rather, a stroke which killed him, at age 67. Coincidentally, his attending physician was Kitty's ex-husband Stewart Harrison, who was called in as a friend of the family. On the other hand, although it seems that Ruth Tolman destroyed many of her personal papers, a little of her correspondence with Oppenheimer is preserved.[11] There are two letters — undated — containing passages that on their face indicate an intimacy between them which could be interpreted as something beyond a close friendship. But who knows? In one of them she writes, "The precious times with you that week and the week before keep going through my mind, over and over, making me thankful but wistful, wishing for more. I was grateful for them, Dear, and as you know, hungry for them too." Later in the letter there is an affectionate reference to her husband. In another she writes about a possible arrangement by which they might be able to see each other. She describes driving to San Diego and seeing, "the long stretch of beach where the sandpipers and gulls played. Oh, Robert, Robert. Soon I shall see you. You and I both know how it will be." She describes the guest list for a party she is planning to give then. Christy is mentioned as one of the guests, but he was not sure that he could remember the occasion. How one puts all of this together I do not know. But to repeat: the affair, if it happened, was not "notorious" and it did not seem to break anyone's heart. Incidentally, Edith Jenkins provides a remarkable insight. She writes, "Oppie was already married to Kitty. I asked Jean Tatlock if she regretted refusing to marry him, and she said yes, she would not have done so had she not been so mixed up. I recall responding perhaps that she perceived him as essentially nonsexual. Jean put her cheek against Margy's (Jenkins's child) and held gently the baby hand that was pulling her hair and said, 'Maybe you're right....'."[12]

Herken's book received a certain amount of notoriety because of its claim that Oppenheimer was actually a member of the Communist Party. If this is true, it means that he lied repeatedly to various security agencies in an entirely unambiguous way — an extremely serious charge. On what is it based? Once again we have Chevalier to thank. During his pre-war Berkeley association with Chevalier, Oppenheimer attended meetings of a small group of left-leaning people — mostly, if not exclusively, Communists. He called this a "coffee klatch." In July of 1964, he received an entirely unexpected letter from Chevalier, with the salutation "Dear Oppe."[13] "Opje" — a Dutch diminutive — means, roughly, "little Op." This is how he was known before the war; "Oppie" came later...

Chevalier begins by saying that "I had not thought ever to write you again. Our ways have parted so utterly." But he had changed his mind. He was writing a memoir — as opposed to a fictionalized novel — in which he intended to discuss his early years with Oppenheimer. But then comes the bomb-shell: "The reason for my writing you is that an important part of the story concerns your and my membership in the same unit of the CP from 1938 to 1942." In other words he is saying that Oppenheimer was a member of the Party in the very years when he was already engaged in working on nuclear weapons. He seems to think that Oppenheimer has nothing "to be ashamed of," and that it would be all right to publish this, unless Oppenheimer had "any objections." In writing this letter it is difficult to tell whether Chevalier is being naive or malicious. Didn't he understand that for decades Oppenheimer had given sworn statements that he had never been a Party member? To have been caught lying about this would not only have destroyed Oppenheimer's reputation, but would have exposed him to serious criminal charges. A few weeks later Oppenheimer replied. He denied in the strongest terms that he had ever been a member of the Party. The matter was turned over to his lawyers, and the accusation was never printed. In short, we are at an impasse. There is no reason, in my view, to believe Chevalier's assertion as opposed to Oppenheimer's denial. The recollections or suspicions of a few aging contemporaries tend to support Chevalier. But unless someone can come up with solid evidence, I simply do not believe either Chevalier or Herken. The reader can study the documents and decide for him or herself.[14]

What I know is untrue is Herken's claim about Oppenheimer's first name. In a footnote on p. 11 of *Brotherhood of the Bomb*, he writes, "The J at the beginning of his name was given to him at birth by his father Julius. The initial did not stand for anything...." The fact that in official FBI documents, and documents involving proofs, Oppenheimer is referred to as "Julius Robert Oppenheimer" might have given Herken pause. But there is no mystery. One needs only go to the Manhattan Bureau of Records and get a copy Oppenheimer's birth certificate. This document, which I have in front of me, says that Oppenheimer was born on April 22, 1904 at 200 West 94th Street in Manhattan — the family residence. It gives his father's age as 33 and his birthplace as Germany. His mother Ella — née Friedman — was born in this country. The name on the birth certificate is Julius Robert Oppenheimer. Why does this matter? I often discussed Oppenheimer with Rabi, who had known him for most of his adult life. His wife had even gone to Ethical Culture school with Oppenheimer. Rabi said again and again that Oppenheimer's problem was identity. One time he told me that Oppenheimer reminded him of a friend who could never decide whether he wanted to join B'nai Brith or the Knights of Columbus. Another time he said that Oppenheimer would have been a much greater physicist if he had studied Yiddish rather than Sanskrit. Oppenheimer was always trying to create identities for himself, and early on he decided that "J. Robert" presented a much better figura than "Julius Robert."[15] I think part of what Lawrence called Oppenheimer's "leftwandering" was just that — a search for identity. It was easy for Oppenheimer to take on an entirely different identity once he became deeply involved with the bomb. He did not find it difficult to support Groves when he tried, unsuccessfully, to make Los Alamos a military base, with everyone in uniform, including Oppenheimer, who would have had the simulated rank of lieutenant colonel. This persona had no trouble telling Groves that his good friend Chevalier was "quite a Red," or that his brother was involved in an espionage attempt. Once the bomb was built he found it easy to become a "smiling public man," who referred to General Marshall as "George." And, after the hearing, there was a new persona — the martyr. I saw this in action on one remarkable occasion. It was a Saturday morning and I was working in my office at the Institute for Advanced Study. Suddenly there was an influx of TV cameras followed by Oppenheimer and the journalist

Howard K. Smith, all heading for our colloquium room. I followed. Oppenheimer said I could stay so long as he did not see me — a very strange request. I was able to hear what ensued. When the cameras were rolling he spoke in that subdued and mesmerizing way about his case and his lack of influence on nuclear affairs. But when the cameras stopped his voice changed and he talked with Smith about the best New York French restaurants and sailing in the Caribbean. Which was the real Oppenheimer? Probably both — or neither.

I promised to tell about Oppenheimer's encounter with Auden and then I will finish with an aside about the memorial service I attended in Princeton not long after Oppenheimer's death from throat cancer on February 18, 1967. When I first came to the Institute in the fall of 1957, as a postdoctoral fellow, I noticed that one of the other visitors was Reinhold Niebuhr. As it happened, Niebuhr was a hero of mine whom I had heard preach several times at Harvard. He was also a friend of my father's. Thus when I met Niebuhr in the Institute cafeteria I introduced myself and received from his wife Ursula a cordial invitation to visit them. Later she confided to me that Niebuhr had recently suffered a heart attack, and that seeing young people cheered him up. One of my visits was just after I had heard Auden, who was visiting Princeton, give a reading. I told the Niebuhrs about this, and was surprised to learn that they were very close friends of Auden — he had even dedicated his poetry collection Nones to Reinhold. They did not realize that Auden was in Princeton. Not long afterwards I found myself on the train to New York accidentally seated next to Auden. I told him about the Niebuhrs and he seemed very interested. I then forgot about it until some days later when I got a phone call from Oppenheimer's secretary asking me to come to his office for a luncheon appointment. Not knowing what to make of this, I was astounded to find assembled in his office Reinhold and Ursula Niebuhr, the British historian Sir Lewyllan Woodword and his wife, Oppenheimer and Kitty, and, of course, Auden. We then marched up to the Institute cafeteria where we were seated at a central table and served. When I walked in I noticed various colleagues such as Freeman Dyson giving me rather fishy looks. I recall the seating arrangement. I was next to Niebuhr, and Ursula was next to him, with Auden at her side. Oppenheimer was seated across the table.

I waited for some sort of epiphany, but the only part of the conversation I can remember was a pointless anecdote Oppenheimer told me about having studied Sanskrit with Arthur Rider in Berkeley. I do not know which of his personae was on display that day, but it seemed to me like a lost opportunity — or nearly. After lunch I took Auden to meet Dyson and the two of them rapidly became absorbed in playing word games on Dyson's blackboard. I never saw Auden again and neither did Oppenheimer. Incidentally, not long ago, some colleagues of mine from the University of Wisconsin sent me a copy of a letter of recommendation that Oppenheimer had written for me as I was leaving the Institute. One of my strong points, he said, was that I had gotten to know the Niebuhrs. Why he thought this would interest the chairman of the physics department I cannot imagine...

The memorial service was something else. It was a beautiful occasion, with talks by Hans Bethe, Henry DeWolf Smyth, and George Kennan. There was also music by the Julliard String Quartet, and a recording of Stravinsky's Requiem Canticles. I was surprised to see on the program that the music had been selected by George Balanchine. As it happened, I encountered Balanchine at the Pilates gymnasium the Monday after the service. I told him that I had not realized he was a friend of Oppenheimer's. He said that he wasn't, and was very surprised when he was asked, apparently by Oppenheimer, to select the music. This was such an odd recollection that a couple of years ago I decided to try to find out if it really happened. After some considerable detective work I found out what had occurred. In the 1950s something called the Congress of Cultural Freedom was formed. Its purpose was to provide a liberal, somewhat left-wing answer to Communist ideology.[16] Oppenheimer joined, as did Nicolas Nabokov — the novelist's cousin. In 1959, there was a meeting of the Congress near Basel, at which Oppenheimer spoke. Among other things he told the delegates that "I find myself profoundly in anguish over the fact that no ethical discourse of any nobility or weight has been addressed to the problem of the new weapons, of the atomic weapons ... what are we to make of a civilization which has always regarded ethics as an essential part of human life, and which has always had in it an articulate, deep, fervent conviction, never perhaps held by the majority, but never absent: a dedication to 'ahimsa,' the Sanskrit word that means 'doing no harm or hurt,' which you find in the teachings of Jesus and Socrates — what are we to think of such a civilization, which has not been able to talk about the prospect of killing almost everybody except in prudential and game-theoretic terms?"[17]

Oppenheimer and Nicolas Nabokov became friends, and in 1967, when Oppenheimer was dying of cancer, Nabokov was living in Princeton. He was working with Balanchine on a ballet, and he must have discussed with Oppenheimer the possibility of having Balanchine select the music. In thinking about this two things struck me. The first was that even in death Oppenheimer had assumed another persona — that of Balanchine. The second was something that neither Nabokov nor Balanchine knew: the Congress of Cultural Freedom had been funded by the CIA.

Acknowledgments

I would like to thank David Cassidy, Robert Christy, Murph and Mildred Goldberger, Gregg Herken, Gerald Holton, Edith Jenkins, and Tom Tombrello for very helpful comments.

References

1. In the Matter of J. Robert Oppenheimer: Transcript of Hearing Before Personnel Security Board and Texts of Principal Documents and Letters, United States Atomic Energy Commission, MIT, Cambridge, MA, paperback edition 1971, p. 470.
2. Reference 1, p. 888.
3. Haakon Chevalier, *Oppenheimer: The Story of a Friendship*, George Braziller, New York, 1965.
4. Reference 3, p. 11.
5. *Robert Oppenheimer, Letters and Recollections*, edited by Alice Kimball Smith and Charles Weiner, Stanford University Press, Palo Alto, CA, 1995, p. 290.
6. Edith A. Jenkins, *Against a Field Sinister: Memoirs and Stories*, City Lights Press, 1991, pp. 24–25.
7. Reference 3, p. 88. The italics are in the original.
8. Reference 3, p. 116.
9. Reference 3, p. 55.
10. It was the "Christy gadget" — a bomb with a solid core of plutonium that was tested at Trinity and dropped on Nagasaki.

11. I am grateful to Gregg Herken for sending me copies of these letters, which are deposited in the Library of Congress.
12. Reference 6, pp. 30–31.
13. The text of this letter, and the next one I refer to, along with other documents, can be found on the Web at http://www.brotherhoodofthebomb.com.
14. David Cassidy informs me that he ran into an analogous problem when he was writing his biography of Heisenberg. Was Heisenberg a member of the Nazi Party? Since the Nazis kept meticulous records of party members, and Heisenberg was not on any of them, the reasonable conclusion is that he did not belong. What is missing in the Oppenheimer case is similar documentary evidence.
15. One should not underestimate the climate of anti-semitism that prevailed at the time Oppenheimer began calling himself "J. Robert" — as he was certainly doing by the mid-1920s. His teacher of physics at Harvard, Percy Bridgman, in a letter of recommendation written to Ernest Rutherford, felt obliged to say, "As appears from his name, Oppenheimer is a Jew, but entirely without the usual qualifications of his race..." Ref. 5, p. 47.
16. A useful history of this institution can be found in *The Liberal Conspiracy*, by Peter Coleman, The Free Press, New York, 1989.
17. Reference 16, p. 121.

10. Unintended Consequences

On May 7, 2008 Gernot Zippe died in Bad Tölz, Germany. He was ninety. I could find only one obituary of him. It was in the URENCO trade magazine — URENCO being the European consortium that enriches uranium commercially with centrifuges. In fact, even though Zippe and I had been in occasional communication by phone and email for a few years before his death, I did not learn about it until some months after the fact. Through no fault of his own, Zippe's centrifuge has been responsible for a very substantial portion of the proliferation of nuclear weapons to places like Iran and probably North Korea. To blame Zippe would be like blaming Einstein for nuclear weapons because he invented the equation $E = mc^2$. Zippe once remarked that, "With a kitchen knife you can peel a potato or kill your neighbor, it's up to governments to use the centrifuge for the benefit of mankind."

When it became clear that Zippe's English was better than my German, all our communications were in English. While he answered many of my questions about the centrifuge, I never could get him to tell me much about his life. As far as I could gather he was Austrian. At least in the 1930s he studied at the University of Vienna. Exactly what he studied and how far he got I did not find out, but my impression was that it was physics and engineering. I always called him "Doctor Zippe" but I have no idea if he ever got a Ph.D. or

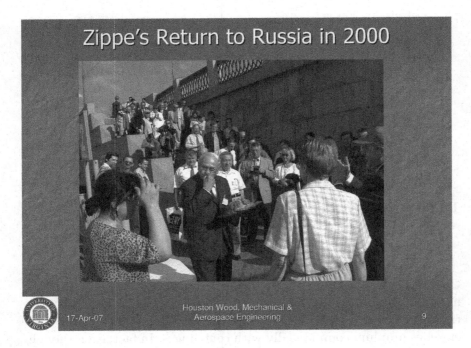

whether this was an honorific. When the war came he served in the Luftwaffe — the German air force as a flight instructor. (He was still flying planes when he was 80.) Somehow he was captured by the Russians and interred in a pretty bleak prison camp. To understand the next step we must introduce the flamboyant character of Manfred von Ardenne.

Unlike Zippe, Ardenne was only too willing to talk about himself. He even wrote an autobiography. The problem was what to believe. In any event he was born in 1907 in Hamburg into a wealthy aristocratic family. By age fifteen he had received his first patent in electronics. By the end of his life he had some 600 patents in various fields. He never completed his formal education and when in 1928 he came into his inheritance he used it to establish a private laboratory on an estate in Berlin. When the war came his laboratory, which by this time was financed by the German post office, began to devote itself to nuclear energy. I have never been able to learn exactly what Ardenne's intentions were. Was he thinking of a bomb? The official German program which had people like Werner Heisenberg attached to it, seemed to look down on the work being done by Ardenne's group as an irrelevance. Actually in some ways Ardenne's people got at least as far, if not further. Two things are worth mentioning.

In 1940 the Dutch-Austrian physicist Fritz Houtermans, who had the unfortunate distinction of being imprisoned in Russia by the NKVD and then being turned over to the Gestapo, found employment in von Ardenne's laboratory. Houtermans soon proposed using element 94 — which had secretly been named "plutonium" in the United States — as a fissile material. He was so worried about the implications of this that he tried to warn people he knew in the United States. The same plutonium suggestion was made at about the same time in the official program by C.F. von Weizsäcker. Having read both proposals, Houtermans' was much more sophisticated. Of more relevance to us, was the effort of von Ardenne's laboratory to separate the isotopes of uranium. Ardenne used an electromagnetic method much like the Calutron that was used at Oak Ridge to make the final separation of the isotopes. I have not been able to learn how much, if any, the Germans actually separated, but when the Russians occupied Berlin they went straight to Ardenne's laboratory and took him and his equipment back to the Soviet Union. Houtermans was long gone.

Ardenne, along with several other scientists, including the Nobelist Gustav Hertz, were installed in a suburb of Sukhumi on the Black Sea. Ardenne was soon approached by Lavrentiy Beria, Stalin's appointed deputy to head the Russian atomic bomb project, to work on the bomb. Ardenne took on the mission of separating uranium isotopes. He split the several German scientists who were there into three groups. Hertz took on separation by gaseous diffusion, a process in which gaseous uranium is forced through tiny pores in a membrane. The lighter isotope U-235 emerges first which is the separation. This method was used in our effort at Oak Ridge. Ardenne took on electromagnetic separation, which is what he had been doing in Germany and a physicist named Max Steenbeck was put in charge of the centrifuges. Steenbeck had been captured by the Russians and put in a grim concentration camp in Poznań. Somehow he managed to communicate his scientific background to the NKVD and soon found himself in Sukhumi with Ardenne. Apparently Zippe followed the same trajectory although he did not want to discuss it. Neither Steenbeck nor Zippe had ever worked on centrifuges and knew nothing about them. Steenbeck took charge of the theoretical work and Zippe the experimental.

Zippe told me what they had available to them when they started. They had some old and not very efficient Russian centrifuges and some articles in American journals. There was a review article by Jesse Beams of the University of Virginia which pretty much summed up the state of the art for gaseous centrifuges — centrifuges in which the spinning material was a gas rather than a liquid — as it existed before the war. All these uranium separation methods use a uranium gas in which uranium is combined with fluorine. Beams had been the first person to use gas centrifuges to actually separate isotopes. The irony is that the uranium centrifuge program that Beams initiated during the war was soon abandoned because the centrifuges were not reliable. Apart from reliability, Zippe and Steenbeck realized that efficiency was crucial. How much power does it take to run a single centrifuge? To separate any significant amount of uranium requires many thousands of centrifuges and thus efficiency is absolutely essential. A good modern centrifuge consumes something like forty watts of power — less than a dim light bulb — and this sort of thing is what Zippe and Steenbeck were trying to achieve. Below is a diagram of what is sometimes called the "Zippe centrifuge" although when he talked with me he always called it the "Russian centrifuge."

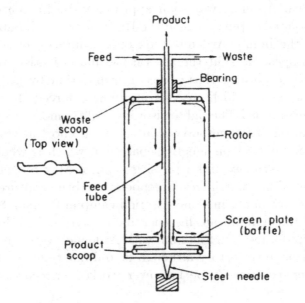

There are a couple of things I would like to call your attention to. The arrows represent the flow of gas. The heavy isotope is flowing down while the lighter one flows up. This double flow is called "counter current" and was invented before the war by the American physicist Harold Urey. At the top, the rotor is held in place by a magnetic bearing which is more or less frictionless while at the bottom it is balanced on a steel needle. Neither of these ideas was original with the German group but they put them together better than anyone had done before. Their best centrifuges were made of aluminum tubes with a diameter of about 100 millimeters. What is essential is the peripheral speed — the speed that a dot attached to the cylinder goes around. This is very important because the separative work that an individual centrifuge can do is approximately proportional to at least the second power of this speed. The best Russian centrifuges had peripheral speeds of about 350 meters a second — faster than the speed of sound in air.

In 1956 Zippe was released and chose to go west. Ardenne and Steenbeck remained in East Germany. Zippe told me that he was not allowed to take any documents with him but that this did not matter since he had the details of the centrifuge in his head. He was sure that the reason that the Russians had let him go was that they had lost interest in the centrifuge program. He learned later that this was wrong — they had a vast centrifuge program — so he was never sure why they let him go. Soon after he returned to Germany he learned that there was a conference on centrifuges in Holland. He went and was surprised to learn that the Russian centrifuge was better than any of the others that were on offer. Among other people he spoke to a Dutch centrifuge designer named Jacob Kistemaker. Kistemaker was so impressed that he changed his whole program so that it followed along the lines of the Russian centrifuge. Zippe became a consultant to Kistemaker. The delicate matter of patents was not raised by the Russians since they did not want to reveal their interest. As it happened there were centrifuge programs in Germany, Holland and England. The German company Degussa, which had an unsavory Nazi past, got into the business and Zippe became a consultant there. Degussa sold some centrifuge technology to Iraq. But in 1964 the Germans formed a state-owned centrifuge company which was in competition with the Dutch and the British. In 1970 the German company was privatized and entered into a consortium agreement with the Dutch and the British and thus URENCO was formed. All the centrifuges that they built were variants of the Russian design. Zippe himself spent the years 1958–1960 at the University of Virginia with Beams designing improved centrifuges then returning to Germany where he was a consultant for industry. When I spoke to him he was living in retirement with his son. The scene is now shifted to Iran.

The Iranian nuclear program began under the Shah in the 1970's. He tried to persuade diaspora Iranians to return and also attempted to hire foreign consultants. Some colleagues of mine had lucrative offers. When he was deposed in 1979 there were a few thousand people working on the program. But the Supreme Leader, who replaced him, after the revolution, the Ayatollah Khomenei, decided that nuclear weapons were "un-Islamic" and reduced the program to a few hundred workers. When he died in 1989 he was replaced by the present Supreme Leader, the Atatollah Khamenei. The president became Akbar Hashemi Rafsanjani. By this time Iran and Iraq were at war and Saddam Hussein was actively trying to produce nuclear weapons, something the Iranians may have known. In any event they became determined to get them for themselves.

They realized that this is something they could not do alone so they turned to the Pakistani nuclear weapons proliferator A.Q. Khan. Khan was more than willing. In a television interview he gave on the 31st of August 2009 in Pakistan he said,

> "Iran was interested in acquiring nuclear technology. Since Iran was an important Muslim country, we wished Iran to acquire this technology. Western countries pressured us unfairly. If Iran succeeds in acquiring nuclear technology, we will be a strong bloc in the region to counter international pressure. Iran's nuclear capability will neutralize Israel's power. We had advised Iran to contact the suppliers and purchase equipment from them."[a]

---

[a]I am grateful to Robert Norris for supplying me with a copy of this interview which is distributed by the Open Source Center.

In the early 1990's with Rafsanjani's support the Iranians made contact. One of Khan's branch offices was in Dubai where he had what amounted to a supermarket in nuclear items with a price list. In October of 1994, for some three million dollars in cash, the Iranians went on a shopping spree. They bought some used P-1 centrifuges which still had traces of highly enriched uranium and they bought plans for the successor P-2. They also bought plans for making metallic uranium hemispheres which when joined form the "pits" of a nuclear weapon. They probably bought plans for the Chinese design of a nuclear weapon which had been given to the Pakistanis in exchange for centrifuge technology. Rafsanjani was complicit in all of this. As far as I can see, nothing has changed his mind.

Here is something he said at Friday prayers in December of 2001. "If a day comes when the world of Islam is duly equipped with the arms Israel has in its possession, the strategy of colonization would face a stalemate because application of an atomic bomb would not leave anything in Israel but the same thing would just produce damages in the Muslim world." The latter part of this statement persuades me that Rafsanjani has no comprehension of what a nuclear weapon is. Israel has hundreds of nuclear weapons. A handful of these would reduce Iranian cities to rubble. On July 17, Rafsanjani spoke again at Friday prayers. What he said about the elections has been widely quoted. But what was overlooked was the following,

> "Our country should be united against all the dangers that threaten us. They have now upped their ransom demands and are coming forward to take away our achievements in the fields of hi-tech and particularly nuclear technology. Of course, God will not give them the opportunity to do so, but they are greedy. My brothers and sisters, first of all, you all know me, I have never wanted to abuse this platform in favour of a particular faction and my remarks have always concerned issues beyond factionalism. I am talking in the same manner today. I am not interested in any factions. In my view, we should all think and find a way that will unite us to take our country forward and save ourselves from these dangerous and bad effects, and the emerging grudges. We should disappoint our enemies so that they would not covet our country."

I have not heard a word from any of the Iranian political figures of any persuasion offering any sort of compromise about the Iranian nuclear program. Where then do things stand?

It appears as I write this that the Iranians now have some 15,000 centrifuges into which they are currently feeding uranium hexafluoride gas. They have produced about 9,000 kilograms of 3.5% enriched uranium. This is the enrichment needed to run a reactor like the one at Bushehr. To run this reactor for a year requires about 28,000 kilograms of LEU and at least three times as much to start it.[b] If this low enriched uranium is really going to be used for this reactor, the Iranians are still far from their goal. Prior to putting uranium hexafluoride gas into the centrifuge one puts the cylinder into which the gas will eventually go under vacuum. This allows the cylinders to turn more rapidly. The Iranians have about nine thousand centrifuges under vacuum. They have also produced about 170 kilograms of 20% enriched uranium which could be converted into the highly enriched uranium needed for a bomb in a matter of months. But the product of this centrifuge enrichment is a gas. This must be reduced to a metal. There is little doubt that the Iranians are quite capable of doing this. They no doubt bought the plans from Khan. A Hiroshima style bomb requires about 60 kilograms of highly enriched uranium. The clock continues to turn.

---

[b]I am very grateful to Richard Garwin for pointing these numbers out.

11. An Unlikely Collaboration

Reprinted with permission from *Physics in Perspective*, **12**, 36 (2010). Copyright Springer

"One proposed design for "Super"[a]

After von Neumann's death in 1957, *Life* magazine revealed that in a 1950 interview he had told an interviewer, "If you say why not bomb them [the Russians] tomorrow I say why not bomb them today? If you say today at five o'clock, I say why not one o'clock?"[b] This same year the German-born British theoretical physicist Klaus Fuchs was arrested in England and confessed to having been a spy for the Soviet Union. I do not know if von Neumann knew this when he made his statement. If so, he may have recalled that in the spring of 1946 when they were both at Los Alamos, he and Fuchs applied for a patent for a version of the hydrogen bomb, after being substantially modified, was actually tested on the 8th of May 1951. The test, known as "Greenhouse George," was done on the island of Eberiru on the Eniwetok atoll in the South Pacific. It was the largest nuclear explosion ever done up to that time. Its yield was some ten times

---

[a]This is the "Probable Value" assigned to a patent filed in the name of John von Neumann and Klaus Fuchs in May of 1946. I am grateful to Carey Sublette for this document.

[b]See http://onion.math.uwaterloo.ca/~hwolkowi/Neumann1.html

greater than the bomb that destroyed Nagasaki on August 9th 1945. Many of the details of the design of that fission bomb had been given to the Russians by Fuchs in earlier 1945 and certainly speeded up their development by years. What I want to explain here is that, while Fuchs turned over the details of his patent with von Neumann for a hydrogen bomb to the Russians in March of 1948, this work seems to have had no effect on their program. It had no explicit affect on the American program either, which is nearly as much of a paradox as the Fuchs–von Neumann collaboration itself. The device tested at Greenhouse George was almost an afterthought. By this time the real hydrogen bomb had been invented by the Polish-born mathematician Stanislaw Ulam and Edward Teller. It was tested on November 1, 1952, also in the South Pacific, and produced a yield about a hundred times greater than Greenhouse George. To understand all of this I need to give a brief history of the development of the hydrogen bomb.

As far as I can tell, the first discussion of how to design a thermonuclear device — a hydrogen bomb — took place in September of 1941 at Columbia University where Enrico Fermi was then a professor and Edward Teller was working on the Manhattan project that was exploring the possibilities of nuclear weapons. As Teller later recalled," It [the hydrogen bomb] started at Columbia, and in this case it was an idea of Enrico Fermi that triggered it. We usually had lunch in the faculty club and as we came back from lunch to Pupin [where the physics department was located] I remember that Fermi stopped just before Pupin and said, "Now if the nuclear bomb works we can reproduce fusion ... the energy source in the Sun, except of course we would not use hydrogen but deuterium [heavy hydrogen whose nucleus has a neutron in addition to the proton which is the nucleus of ordinary hydrogen] where the cross-section [which determines the rate of the reactions] is very much bigger." I gave it some thought and practically a week later [on another walk] I proved to Fermi that it was a bum idea."[c] Teller then moved to Chicago but still obsessed by Fermi's idea which he could not quite convince himself

---

[c]This is from an interview that Teller gave to Jay Keyworth in September of 1979. I thank Richard Garwin for supplying a transcript of this interview.

was wrong. The next development came in the summer of 1942 in the compartment of a train which Hans Bethe shared with Teller. They were on their way to San Francisco to attend a meeting that had been organized by Robert Oppenheimer to study the design of atomic weapons. Los Alamos was not created until the following spring. Bethe was then a professor at Cornell while Teller was at the University of Chicago. So Bethe stopped off in Chicago to pick up Teller and to get a look at the reactor Enrico Fermi and his group were then constructing. After having seen their work, Bethe became convinced that the reactor would work and that a nuclear weapon would also probably work. In an interview I had with Bethe many years ago this is what he told me:

> "We had a compartment on the train to California, so we could talk freely. Teller told me about the idea of making plutonium in the reactor and using the plutonium in a nuclear weapon. Teller told me that the fission bomb [The Hiroshima and Nagasaki bombs were fission bombs.] was all well and good and, essentially, was now a sure thing. In reality, the work had hardly begun. Teller likes to jump to conclusions. He said that what we really should think about was the possibility of igniting deuterium by a fission weapon — the hydrogen bomb. Well, the whole thing was far more difficult than we thought then. About three-quarters of our time that summer was occupied with thinking about the possibility of a hydrogen super-weapon. We encountered one difficulty after another, and came up with one solution after another — but the difficulties were clearly in the majority. My wife knew vaguely what we were talking about, and on a walk in the mountains in Yosemite National Park she asked me to consider carefully whether I really wanted to continue to work on this. Finally, I decided to do it. It was clear that the super bomb, especially, was a terrible thing. But the fission bomb had to be done, because the Germans were presumably doing it."[d]

To understand the "difficulties" and how the von Neumann–Fuchs invention addressed some of them I need to review a bit of the physics of nuclear weapons.

In principle, nuclear weapons are of two types — fission and fusion. In practice, most weapons are a mixture. Fission weapons, like the ones that destroyed Hiroshima and Nagasaki, derive their

---

[d]See Jeremy Bernstein, *Hans Bethe: Prophet of Energy*, Basic Books, New York, 1980, p. 73.

energy from the process of nuclear fission. This happens when a heavy nucleus like one of the isotopes[e] of uranium or plutonium breaks up into a pair of lighter elements. This can happen spontaneously or it can be induced when the parent nucleus absorbs an ambient neutron. When this happens a "compound nucleus" is formed, generally in a state of high excitation. This excitation causes the compound nucleus, which can be thought of as analogous to a liquid drop, to vibrate very rapidly and ultimately beak up into fission fragments which are nuclei somewhere in the middle of the periodic table. For example, the first fission of uranium, which was observed at the end of 1938 by the German chemists Otto Hahn and Fritz Strassmann, was into barium and krypton. These fission fragments are generally what are known as "neutron rich." Their nuclei contain more neutrons than protons. To move towards stability, where the number of neutrons and protons approach equality, the fragments almost immediately shed neutrons — on the average something like three. These neutrons can cause other nuclei to fission creating a chain reaction. When enough material is present, this chain reaction can run away in microseconds producing an explosion. There is energy created in fission because the final products are less massive than their parents. The mass difference produces energy according to Einstein's formula $E = mc^2$. This energy is carried off primarily in the energy of motion of the fission fragments.

While fission involves heavy elements such as uranium or plutonium, fusion involves light elements such as the isotopes of hydrogen. There are three. The nucleus of ordinary hydrogen consists of one proton. The nucleus of "heavy" hydrogen — the deuteron — consists of one proton and one neutron while the nucleus of "superheavy" hydrogen — the triton — consists of two neutrons and one proton. We shall follow the usual notation when we call the proton P, the deuteron D and the triton T. A typical and very important fusion reaction occurs when two deuterons fuse to produce one triton and one proton. Symbolically we can write this as $D + D \rightarrow T + P$.

---

[e]Nuclei are isotopes of each other if they consist of the same number of protons but different numbers of neutrons.

This generates energy because the triton and the proton have less mass than the two deuterons. This mass-energy is largely taken off by the motion energy of the proton. The first difficulty we have to deal with is that according to classical physics this interaction is impossible.

The reason for this is that each deuteron carries a positive electric charge and like charges repel. These like charges set up a barrier which classical physics tells us cannot be penetrated. It would violate the conservation of energy. But in quantum mechanics energy conservation can be violated if it happens in a sufficiently short time. This is one of Heisenberg's uncertainty principles. Therefore there is a probability that the two deuterons can penetrate the barrier. Once they do, the strong nuclear force takes over and the fusion reaction is completed. We know that this sort of fusion produces the energy in the Sun and the other stars. The reactions are different there but it is the same principle. It works for the Sun because its very high central temperature — millions of degrees — corresponds to very energetic nuclei which aids the fusion. So to make deuterium "ignite" — to fuse — you need to produce temperatures like that of the Sun. Such temperatures are produced by an atomic bomb. This is what Bethe was referring to.

The first question we want to answer is why are fusion bombs so powerful as compared to fission bombs. The most energetic fusion reaction known involving hydrogen isotopes is the fusing of a deuteron and a triton to produce a stable isotope of helium plus a neutron. Symbolically $D + T \rightarrow He^4 + N$. The superscript 4 refers to the fact that this isotope of helium has two neutrons and two protons in its nucleus. This reaction generates something like a tenth of the energy of a typical fission reaction. Hence the puzzle. The answer is that the mass of the deuteron and triton is about a fiftieth of the mass of, say, uranium. Hence a gram of this uranium contains about a fiftieth of the atoms compared to a gram of deuterons and tritons. This compensates for the difference in energy. You get more energy per gram in fusion than you get in fission because there are more nuclei. To put the matter more specifically, if you were to fission one kilogram of uranium it would produce an energy equivalent to twenty thousand **tons** of TNT, but if you were to fuse a kilogram of

D and T it would produce about 80 thousand tons of TNT equivalent energy.

I have to confess that when I interviewed Bethe on his work on the early versions of the hydrogen bomb — the "classical Super" as it was called — I knew next to nothing about the physics of nuclear weapons. This was despite the fact that I had spent the summer of 1957 as an intern at Los Alamos and had witnessed two nuclear explosions in the desert in Nevada. There was a strict "need to know" at Los Alamos and since I was not working on weapons no one told me anything. When Bethe told me about the idea of "igniting deuterium" with a fission bomb the image I had was of dropping a lighted match in a container of gasoline. The image I should have had was trying to ignite a log with a match. I would probably be able to ignite a small patch of the log near the match, but then it would cool off before the fire could propagate throughout the log. This is what calculations showed would happen with the Classical Super. The heat of the fission bomb would ionize whatever atoms were around, meaning that atomic electrons would be torn off. When these electrons interacted with the nuclei they would become accelerated and this produces radiation which leaks out and cools off the configuration. In short, the fission bomb might ignite the deuterium but this ignition would not propagate. Teller proposed one configuration after another and none of them seemed to work. I.I. Rabi who was a witness to all of this told me that Teller reminded him of a man who used to come and see him about a perpetual motion machine he had invented. Rabi would patiently explain to him why it could not work. The man would thank him and in a few days come back with the design of a new perpetual motion machine.

In the late summer of 1945 Fermi gave lectures on the Classical Super. He had a gift for going to the essence of everything and presenting the results in the simplest terms possible. Fermi discussed how fusion is produced in the Classical Super and how the energy is lost. At the end he summed up by saying that "So far all schemes for initiation of the super are rather vague." This I think was being polite. Fuchs was at the lectures and he turned the notes over to the Russians. They were translated into Russian and in January of 1946 presented to a special committee of experts presided over by Lavrenty

Beriya who had been put in charge by Stalin of the Russian bomb program.[f] Probably the most important thing they learned was that the Americans were working on the hydrogen bomb and considering the fusion of deuterium and tritium. But in April of 1946 Teller organized a three-day conference and Fuchs and von Neumann attempted to explain to some people at the conference what their new idea was. No one seemed to be interested. One thing that interests me is how this unlikely pair ever collaborated in the first place.

Neumann János Lajos — using the Hungarian name order — was born in Budapest in 1903. His father was a non-practicing Jewish lawyer who worked for a bank. It was evident from the beginning that the young von Neumann — the "von" had been awarded to his father — was a prodigy. His father was concerned that the practice of mathematics might not be a practical way of earning a living so at the age of 22 von Neumann took PhD's in both mathematics and chemical engineering. After his father's death in 1929 the von Neumanns moved to the United States. Such was his mathematical reputation that he became one of the first faculty members — joining Einstein — at the newly created Institute for Advanced Study in Princeton. He spent the rest of his career there. Von Neumann was a very social individual. He loved parties, food and drink and ribald limericks. Einstein referred to him somewhat disdainfully as a *denktier* — a thinking animal. With the war he became a consultant at Los Alamos where his chemical engineering background came in handy. He helped to design the explosive lenses that were used to implode the sphere of plutonium used in the Nagasaki bomb. One of the things he did after the war was to create the logical architecture of the computer. When I was an undergraduate at Harvard he came to the university to give lectures on the computer and the brain. They were the best lectures I have ever heard on anything — like mental champagne. After one of them I found myself walking in Harvard square and looked up to see von Neumann. Thinking, correctly as it happened, that it would be the only chance I would

---

[f]I am grateful to German A. Goncharov for this information and for other details concerning the Russian program.

have to ask him a question. I asked, "Professor von Neumann, will the computer ever replace the human mathematician?" He studied me and then responded, "Sonny, don't worry about it."

Klaus Emil Julius Fuchs, the third of four children of a Lutheran pastor, was born in 1911 in Rüsselsheim, Germany. He became a member of the German Communist Party in 1932 when he was a student at Kiel University. When the Nazi's took over it was clear that his life was in danger so he emigrated first to France and then to England where in 1937 he took his PhD at the University of Bristol. It was here that he first met Bethe who had also left Germany. Fuchs then went on to Edinburgh to work with Max Born. When the war broke out Fuchs was detained as an enemy alien and sent to Canada where he used some of his time to give instructive physics lectures to some of his fellow detainees. But Born managed to get him released so he could return to England. By this time Rudolf — later Sir Rudolf Peierls — was working on nuclear weapons. He invited Fuchs to join him at the University of Birmingham where he was then teaching. The Peierls's often boarded physicists in their home. Bethe had lived with them for awhile and now they boarded Fuchs. As I can testify from personal experience, Peierls's Russian-born wife Genia was a force of nature. Fuchs was a mystery to her. He did not say anything unless asked a direct question. He reminded her of one of those machines that played music only when you put a coin in.

Fuchs and Peierls did some important work together. For example they wrote a fundamental paper on the theory of isotope separation. To use uranium for a bomb you must separate the two isotopes uranium-235 and uranium-238 and they discussed, among other things, how to use centrifuges for this purpose. Some of the British group were invited to come to the United States to continue their work. When Fuchs received such an invitation he attempted to decline it on the grounds that what he was doing in England was more important. When the Germans invaded Russia, Fuchs decided that it was necessary for the Russians to get the bomb so he began transmitting information. He certainly must have transmitted the work he had done with Peierls. His British spy connections were well-established and he did not want to lose them. But in any event, Fuchs agreed to come to the United States and in August of 1944 he was at Los Alamos as a member of the Theoretical Division of which Bethe was the head. At Los Alamos he was very well-liked. Since he was non-social he volunteered to act as a baby sitter for his more sociable colleagues who wanted to attend the many parties. He was extremely competent and had an almost photographic memory. Next to Oppenheimer he may have known more about all the activities of the laboratory than anyone. He was constantly transmitting this information to the Russians. In June of 1946 he returned to England where he eventually headed the theoretical program to make the British bomb. For a while he stopped his espionage activities but then in 1947 he presented himself to the Russian embassy. This kind of self-recruitment is not how spies were supposed to operate so at first the Russians were suspicious. But they decided that he was sincere and gave him as a control Alexandre Feklisov, a master spy who had handled the Rosenbergs. When Feklisov offered Fuchs money, Fuchs was outraged. He only took money in 1949, the last year of his espionage activity, when a brother contracted tuberculosis and sought treatment in an expensive clinic in Switzerland.

Given this background one can appreciate how strange the Fuchs–von Neumann collaboration was. Who started it and who contributed what? There are very few clues. An interesting one is the patent application "One proposed design for the 'Super'" that was filed at Los Alamos in May of 1946. The version I have is essentially totally redacted. The proposed claim reads, "The combination with a device for initiating a thermo-nuclear reaction which employs a quantity of fissile material adaptable to sustain a neutron induced divergent chain reaction, a massive quantity of material in which a thermo-nuclear reaction be maintained." It carries essentially zero information. But the significant thing is that von Neumann is given as the first author. This presumably means that he instigated the collaboration. A second clue comes from an FBI report by an agent Robert Lamphere who interviewed Fuchs in jail in England in 1950. Lamphere was not a physicist so he did not question Fuchs in as much detail as he might have. In any event, the FBI was more interested in Fuchs's contacts than in what he actually transmitted. To understand Lamphere's question and Fuchs's answer I need to say something additional about the physics.

Fusion is what is known as a "binary" process. This means that the fusing nuclei react in pairs. What this implies is that the rate of the reaction is proportional to the product of the number of each of the fusing nuclei in a unit volume. If the two fusing nuclei are identical as in a D-D fusion this product is the square of the number densities. How can we increase this density? We can shrink the volume. We will then have the same number of particles but in a smaller volume so the density is correspondingly higher. From this it follows that to increase the fusion rate one method is to compress the volume in which the fusing particles are contained. In the D-D case if you compress the volume by a factor of ten the rate increases by a factor of 100. When Lamphere asked Fuchs who thought of this, Fuchs said that he did. "[Fuchs] stated laughingly that this was his Fuchs' suggestion, and that he did not furnish information concerning the ignition of the super bomb by the implosion process."[g] The last part of this quote is quite incomprehensible to me because what Fuchs did was to precisely turn over this information. What role von Neumann played in the invention is not clear. To understand what Fuchs turned over I call your attention to the diagram below. It first appeared in Greg Herken's book *Brotherhood of the Bomb*.[h] He thanks Joe Albright and Marcia Kunstel who were correspondents in Moscow where they acquired it. The version in Herken's book has Russian captions which are presumably translations of Fuchs' captions. Carey Sublette, who supplied this version, had the captions re-translated into English.

The box stands for a fission device. Von Neumann and Fuchs specified a gun assembly weapon of the kind that flattened Hiroshima. It consists of a target made of highly enriched uranium and a projectile also made of highly enriched uranium. When the target is joined with the projectile by firing one against the other a super-critical mass is formed and the explosive chain reaction follows. But the momentum of the projectile acts as a ram which produces the first compression of the capsule, made in this instance from beryllium oxide. The capsule

---

[g]See *The Brotherhood of the Bomb* by Gregg Harken, Henry Holt and Company, New York, 2002, 374.

[h]Op. cit.

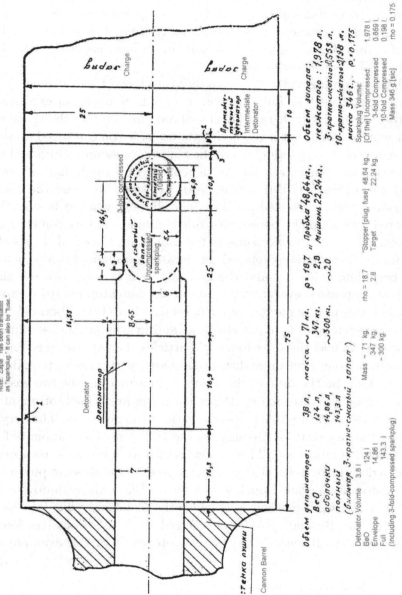

here contains the D-T mixture which here is in liquid form. The larger of the two spheres shown in the diagram represents this brute force mechanical compression. It is probably optimistic to imagine such a perfect geometrical shape resulting from such a collision. In the diagram this is supposed to result in an increase in the density of the D-T by a factor of three.

There is no ingenuity here — only brute force. It is the next step that is ingenious. The initial radiant energy from a nuclear explosion is given off largely in X-rays. This radiation moves with the speed of light which means that if you can make use of it you can speed up the next stage of the explosion where the real compression is supposed to occur. It has been known for over a century that the radiation itself produces a pressure. But this is much too small to cause the compression we are interested in. But the radiation, when it is suitably directed, heats up the beryllium oxide container and its contents to a point where all the electrons in the atoms are liberated. In short, the matter is completely ionized. In the von Neumann–Fuchs scheme the beryllium oxide gas and the D-T gas — it is a gas because the heat has vaporized everything — are at the same temperature. Now we can count up the number of particles that have been liberated per atom. Oxygen gives up eight electrons and beryllium four, so counting the two nuclei there are fourteen particles. The hydrogen isotopes each give up one electron and then there are the two nuclei to make a total of four. So the ratio of the number of particles in the two gasses is something like three to one. But when gasses are in equilibrium at a common temperature this is also the ratio of the pressures. The beryllium oxide gas exerts a pressure on the D-T gas which accounts for the second compression. This is now known as "ionization compression" a descriptive term that was introduced by the Russian physicist and historian German Goncharov, who worked with Sakharov, who invented the same method independently. Goncharov has informed me that the Russian physicists referred to it as "sakharization," a pun on the Russian word for "sweetened."[i] In the example of

---

[i]I thank Norman Dombey for pointing this out. "Sakhar" is the Russian word for "sugar."

the above diagram the overall compression is by a factor of about ten. The notions conveyed in the diagram were tested in the Green House George explosion although that device was very substantially modified from the original von Neumann–Fuchs design. People who were present at the time never heard this design mentioned.[j] Von Neumann was a consultant and he must have had something like this in mind although given Fuch's outing he might well have been reluctant to bring up the name of his former collaborator. A completely different form of radiation compression is used in the Ulam–Teller hydrogen bomb which bears no real resemblance to the von Neumann–Fuchs design. I will come back to this below.

While the Greenhouse George test showed that the Fuchs–von Neumann design would ignite the fusible D-T, once ignited, the reaction still would not propagate. An equilibrium condition had been established after the ignition and the heat that was produced was radiated away very rapidly. In short, everyone was stuck. No one knew how to make a hydrogen bomb. Enter Stanislaw Ulam. Ulam was born in 1909 in Lwōw which, after the First World War, became part of Poland. Like many mathematicians his abilities manifested themselves early. Indeed in 1938 he was invited to become a Junior Fellow at Harvard. Arthur Schlesinger Jr. was a Junior Fellow at the same time. But in 1939 he returned to Poland escaping with his brother Adam just before the war. The rest of his family perished in the Holocaust. Adam eventually became a professor of government at Harvard and Ulam became a professor at the University of Wisconsin. Not long after Los Alamos was founded, von Neumann, whom Ulam had gotten to know in 1935 when he was a visitor at the Institute for Advanced Study, recruited Ulam for work on a project — the atom bomb — which von Neumann was not allowed to say anything about beyond the fact that it was interesting and important. Ulam was instructed to proceed to a railway stop near Santa Fe. He sent his wife to the library to take out an atlas of New Mexico. When he looked at the list of recent borrowers he

---

[j]I thank Kenneth Ford, Arnold Kramish and Herbert York who were present for their comments.

noticed that they were physicists who had disappeared from the department.

It is often the case in my experience that when you ask a mathematician about some problem you can't solve he, or she, shows that it is equivalent to another problem which they also cannot solve. Ulam was not like that. He solved problems. One of the things he did during the war was to invent the so-called Monte Carlo method which is used for making approximate analyses of problems which cannot be solved exactly. After the war Ulam decided to stay at Los Alamos. It was inevitable, especially after Truman's decision, that he would turn his attention to the hydrogen bomb. In 1950 he and a younger colleague C.J. Everett decided that they would make an independent analysis of the Classical Super. By June they had concluded that the situation was more hopeless than anyone had imagined. But the exercise gave Ulam a greater mastery of the physics than anyone else.

Here things stood until December when Ulam had a brainstorm. It goes under the rubric "super compression." Suppose you could use the energy of the primary fission bomb directly to compress the fusion fuel.[k] Hypothetically there would be enough energy to compress the fuel not to the factor of ten in Fuchs–von Neumann scenario but to say a factor of eighty. This could compress the fuel — for example deuterium — and everything surrounding it to such a high density that it would become opaque to the radiation being produced in the interior. It would act like a sort of shroud. It would not matter if equilibrium was reached because the radiation would be contained in the interior and would not cool the fusible elements. Ulam's idea was to produce this compression using the detritus from the primary fission device — the neutrons and the fission fragments. In late January he went with this scheme to Carson Mark, a Canadian who had come to Los Alamos during the war and had stayed on to become head of the theoretical physics division. Mark was busy with Greenhouse George and after listening for an hour told Ulam to talk to Teller. Having known Ulam, who died in 1984, I imagine that Mark must

---

[k]It is likely that Ulam first wanted to apply super compression to an ordinary fission bomb to improve its efficiency. But then he realized how it was relevant to the Super.

have heard any number of Ulam's fanciful speculations of which he had at least one a day. It should also be noted that at the time there was no nuclear device that was small enough — a "bomb in a box" — to be used this way.[1]

Ulam and Teller did not like each other to put it mildly. Both men had outsized egos and it could not have helped that Ulam had recently shown that for all practical purposes, Teller's classical Super was dead. That apart Teller had a "theorem" which had persuaded him that no amount of compression, super or otherwise, could help. Teller of course understood that when you compress the fusible elements that increases the reaction rate of the fusions. But he argued that the rates of all the processes by which energy is dissipated by radiation are increased by exactly the same amount — something that is known as "scaling." It turned out that he had left out a crucial process that did not have this property. You could indeed beat the system and compression would work.[m] Once Teller understood this he embraced the idea enthusiastically, so enthusiastically that in his later accounts he more or less claimed to have invented it. What he did invent was the method of compression using radiation which was a vast improvement over what Ulam was proposing. This radiation compression is totally different from the von Neumann–Fuchs idea. Whether he had their proposal somewhere in the back of his mind who can say. In any event, it worked, and we had, for better or worse, the Ulam–Teller hydrogen bomb.

The Russians were alarmed by the Truman edict and began working furiously on their own hydrogen bomb. The first design they tested was not really a hydrogen bomb although it did involve fusion. They called it the *Sloika* a kind of Russian layer cake. In the center

---

[1]Ken Ford has informed me that he had an interview with Carson Mark who made this point.

[m] "Bremsstrahlung" — braking radiation — is a process in which an electron that is been accelerated or decelerated by its interaction with a nucleus emits radiation. But there is inverse Bremsstrahling in which a radiation quantum is absorbed by the electron. This process takes place when a nucleus is present and hence there are three bodies involved and Teller's theorem is evaded. This process contributes significantly to the opacity. I thank Carey Sublette and Ken Ford for comments on this.

was a fission device which was surrounded by spherical shells — layers — of alternating fission and fusion fuels. When the primary was exploded this set off a sequence of secondary events involving both fission and fusion. About the same time as the Americans they hit on the idea of super compression. Who exactly was responsible and how this came about I do not know. However I am quite sure that it had nothing to do with espionage. The Russians have freely admitted that Fuchs gave them the fission bomb, but they have adamantly denied that anyone gave them the hydrogen bomb. German Goncharov, who was there, has informed me that the government minister who was then in charge of the program Andrei Malyshev opposed working on it in favor of the *Sloika*. It was not until 1954 that he changed his mind. The Russians tested their first hydrogen bomb in 1955, the British in 1957, the Chinese in 1967 and the French in 1968.

On November 22, 1955 Goncharov witnessed the first Russian hydrogen bomb test. On the fiftieth anniversary of this test Goncharov published an article with the curious title "The extraordinarily beautiful physical principle of thermonuclear charge design."[n] In it he discusses the role of Fuchs as the Russians perceived it. He tells us that in September of 1945 Fuchs transmitted information about the Classical Super including a diagram which he presents in his article. The effect of this was to alert the Russians that the Americans were working on a hydrogen bomb and that they had better look into the matter. But the Russians, also using material supplied by Fuchs, were busy trying to duplicate the plutonium fission weapon. They made the first successful test of it in August of 1949. In the meanwhile Fuchs had delivered the von Neumann–Fuchs invention via his London contact Alexander Feklisov. Goncharov informs us that this was immediately given to Beriya. He called in two Russian physicists to analyze this new information. It is important to understand that only a miniscule number of Russian scientists were allowed to see any espionage data. Some of it was kept from people who really needed it. Why Beriya took this position I do not

---

[n] *Physics-Uspekhi* **48** (11) 1187–1196 (2005).

know. One of the people who was allowed to see these reports was Yuri Khariton — a leader in the Russian program. He summarized the von-Neumann–Fuchs invention without really understanding it. In fact the Russians did not understand it until 1954 when Sakharov and others designed the real hydrogen bomb. It is part of the irony of this story that the unlikely collaborators John von Neumann and Klaus Fuchs in 1946 produced a brilliant invention that could have changed the whole course of the development of the hydrogen bomb, but was not fully understood until after the bomb had been successfully made. However it would have been better for everyone if it had never been made at all.

12. A Nuclear Supermarket

> "'Oxygen' is basic to life, and one does not debate its desirability, the nuclear "deterrence" has assumed that life-saving property for Pakistan."
>
> General Mirza Aslam Beg, Chief of Staff of the
> Pakistan Army from 1988 to 1991.[a]

> "Dr. A.Q. Khan and his scientists have given the country a credible deterrent for a paltry sum of money. What they have in their accounts is what I call gold dust — they have not taken the government's money. If a scientist is given 10 million dollars to get the equipment, how would he do it? He will not carry the money in his bag. He will put the money in a foreign bank account in somebody's name. The money lies in the account for some time, and the mark-up that it fetches may probably have gone into his account. It is a fringe benefit."
>
> General M.A. Beg[b]

In the spring of 1969, I received word that my appointment as a Ford Foundation Visiting Professor of Physics at the University of Islamabad had been cleared by the Government of Pakistan. I had actually gotten the appointment several months earlier but there had been a not entirely peaceful change of governments in Pakistan. Ayub Khan was forced to resign and had been replaced by Yahya Khan.

---

[a]This is quoted in *Shopping for Bombs*, by Gordon Corera, Oxford Press, New York (2006), p. 74.

[b]Corera, op. cit. p. 146.

("Khan" is a common Pathan name which was sometimes taken by non-Muslims who rebelled against the British.) Yahya had declared martial law and was cracking down on the universities which is why my clearance had been delayed. Now that I was actually going I gave some thought as to how I was going to get there. The obvious answer was to fly, but that seemed like a missed opportunity. I had read some accounts of people driving from Europe to India. There was even an extraordinary account by one Dervla Murphy who, in 1963, had ridden her bike from Ireland to India. In her book "Full Tilt"[c] she says that since she was traveling alone, when she arrived in a town at night, her first action was to go to the local police station and ask to sleep there. I decided that I was going to drive.

This required several steps. First there was the choice of vehicle. That was fairly obvious. There existed a converted Land Rover called a Dormobile which slept four persons and had a gas stove. It preserved all the rugged features of a Land Rover. The Ford Foundation people were somewhat surprised by my choice of transportation but agreed to contribute what they would have spent on an plane ticket towards the purchase of the Land Rover. The rest of the money was loaned to me by the New Yorker on the condition that after I sold the Land Rover I would pay it back. The fact that William Shawn was willing to do this had to do with my choice of traveling companions. These were the Chamonix mountain guide Claude Jaccoux and his then wife Michele. The three of us had spent some months together in Nepal two years earlier and Shawn had liked the article that I had written about it. The Ford Foundation stipulation was that I appear in Islamabad by the first of October when classes began at the university. Thus it was that the three of us left Chamonix on the fifth of September in the Land Rover, headed to Pakistan.

We went through the fairly recently completed Mont Blanc tunnel into Italy, drove to Venice, and then Trieste where we stopped. I wanted to visit an old friend, Abdus Salam. Salam had been born in 1926, in a small town in Pakistan. He was a mathematics and physics prodigy and after getting his undergraduate degree at the

---

[c]In 1987 it was reissued as a paperback by Overlook TP.

Government College at the University of the Punjab he won a scholarship to Cambridge, where he took his PhD in 1951. His idea was to return to Pakistan to try to create a center for theoretical physics. He found that he simply could not get the support to do this so he returned to Britain where, in 1957, he became a professor at Imperial College in London. In 1964 he helped to found what is now known as the Abdus Salam International Centre for Theoretical Physics in Trieste whose purpose was to allow people from developing countries such as Pakistan to work on cutting edge physics. For his own physics he shared the Nobel Prize in 1979 with Sheldon Glashow and Steven Weinberg. He was also a member of Pakistan Atomic Energy Commission which means he must have had detailed knowledge of the Pakistan nuclear weapons program. He died in 1996 and I never had a chance to ask him. He was pleased to see us and pleased that I was going to teach in Pakistan, but our method of getting there he found barking mad.

Our route continued through Yugoslavia, into Greece, then Turkey, past Mount Ararat, and into Iran. In eastern Iran we encountered several hundred miles of unpaved road and had our only flat tire. We had brought two sets of spare wheels. Then we entered Afghanistan and were stunned to find a veritable unused superhighway which was the result of an American–Russian competition to build roads. We spent a little time in Kabul including a side trip to see the giant Buddhas in Bamiyan which were later destroyed by the Taliban. Then we went over the Khyber pass into Pakistan. I knew a good deal about the history of the Khyber pass and was surprised that crossing it in an automobile was not very difficult. Once in Pakistan we headed for Rawalpindi which is the old city that is a twin to the then new Pakistan capital, Islamabad. It was the 30th of September. I made contact with my Pakistani hosts who were very apologetic. Because of the student unrest of the previous spring the university was going to be closed down for a month. They had tried to notify me but, of course, I was en route. This was, in fact, wonderful news. We spent a month exploring the Northwest Frontier from Skardu to Chitral, even doing some treks close to the Afghan border. These must now be the infiltration routes of the Taliban. In the first week in November my abbreviated course began.

The teaching was very light so that I was free to do a good deal of traveling around the country. My hosts arranged a visit to the KANUPP nuclear reactor on the Arabian sea coast near Karachi. The reactor was then under construction. To get some insight into the Pakistan nuclear program, and to these programs in general, it is useful to present a little primer as to how a reactor works. Schematically there are four components. There are the fuel rods largely composed of uranium in which the fission reactions take place. There is a coolant which cools off the fuel rods to keep them from melting down. There is a heat exchanger which takes the heat generated by the fission reactions and converts it into steam which is used to power the turbines that generate electricity. Metaphorically speaking, a reactor is like a superannuated kettle in which fission energy is converted into the energy of steam. Finally there is something called a "moderator" which requires further explanation that I will now give.

For purposes of discussing fission, a heavy nucleus like that of uranium can be thought of as a liquid drop. The collective behavior of the dozens of neutrons and protons in such a nucleus resembles to a degree the collective behavior of the molecules in a liquid drop. When such a nucleus is bombarded by a neutron the "drop" is agitated and if the energetic conditions are right it splits into "droplets." The larger droplets are nuclei of lighter elements than uranium. In the first fission reaction that was observed in December of 1938, by the German radio chemists Otto Hahn and Fritz Strassmann, the "droplets" were nuclei of barium and krypton. By atomic standards these fission fragments are produced with a high kinetic energy. It is this kinetic energy that is converted into heat. These fission fragments shed neutrons and if more than two are produced in this process then they in turn will generate new fissions which produces a chain reaction. As I mentioned, the fission is started by a neutron incident on the "drop." What was unexpected, until Enrico Fermi discovered it in 1934, was that the slower these neutrons are the more readily do they produce nuclear reactions like fission. This is a quantum mechanical effect. Neutrons do not behave like baseballs thrown at a window. They have a wave nature. The role of the moderator is to slow the neutrons. This happens because the neutrons collide with the nuclei of the moderator and give up some of their momentum in the collision.

The ideal moderator would be the nucleus of hydrogen — the proton — since it and the neutron have about the same mass so the proton can recoil like a struck billiard ball and take away a large fraction of the neutron's momentum. Hence ordinary water, whose molecule consists of two hydrogen atoms and an oxygen atom, would seem like the most desirable moderator. Indeed, the water could both cool the fuel rods and be transformed into steam. There is a catch. There is always a catch. Some of the neutrons will combine with protons to make the nuclei of "heavy hydrogen" which consists of two neutrons and a proton. These neutrons are then no longer available to the chain reaction. To compensate, in the so-called "light water" reactors the uranium in the fuel rods is "enriched." Uranium that comes from a mine consists predominantly of two isotopes Uranium-238 and Uranium-235. Over ninety nine percent of natural uranium is Uranium-238 whose nucleus has 92 protons and 146 neutrons. Uranium-235 nuclei have three fewer neutrons. As it happens, Uranium-235 is much more readily fissioned than Uranium-238 so to compensate for the neutron absorption problem with the light water moderator, in such a reactor the uranium is enriched so that its Uranium-235 content is about 3.5 percent. This means that light water reactor technology inevitably gets mixed up with uranium enrichment and the possibility of enriching the uranium so that it can be used in nuclear weapons.

There is another solution which is to use "heavy water" in which ordinary hydrogen is replaced by heavy hydrogen. The nucleus of heavy hydrogen — the "deuteron" — is not that much heavier than the proton so it is a very effective moderator. In fact it is so effective that unenriched uranium can be used as the fuel. Thus the problem of uranium enrichment is replaced by the much simpler problem of extracting heavy water from the natural water that occurs in the ocean or a lake. The first people to make such a reactor were the Canadians. They began work on designing what is known as the CANDU — CANadian-Deuterium-Uranium — power reactor in the late 1950s. The first of these reactors to generate electricity went into service in Rolphton, Ontario in 1962. When the Pakistanis wanted to build their first power reactor — the KANUPP — they hired the Canadians who supplied everything including the heavy water. The KANUPP went into service in 1972 and ran without serious incident for its thirty year estimated lifespan. In 2002 it was shut down so that it could be modernized to extend its life time. During the entire era of its operation it was an open site subject to inspections. So far as I know, no plutonium — an inevitable consequence of any reactor operation — was ever re-processed. In short, the KANUPP reactor seemed to be a model of how such a facility should be operated. It is now a restricted site.

On the other end of the spectrum is the reactor the Pakistanis built with the aid of the Chinese in Khusab in the Punjab. It also is a CANDU type reactor and apparently the Chinese supplied the heavy water for the moderator. The reactor went into operation in April of 1998. From the beginning all of its activities have been secret. From the evidence that has been collected, it is clear that the principle purpose of this reactor has been to produce plutonium. It apparently has also been used for the production of super-heavy hydrogen-tritium whose atomic nucleus has two neutrons and a proton. This is significant because tritium can be used to make either hydrogen bombs or so-called "boosted" atomic bombs. The idea is that under the conditions created by a primary nuclear explosion the tritium and deuterium nuclei can "fuse." This fusion produces a nucleus of helium and gives off energy. But the most important thing that it does is to produce a neutron. Hence in such a weapon a burst of neutrons is produced which greatly enhance the fission chain reaction. The Pakistanis acquired a tritium purification plant from Germany in 1987. There are also plutonium reprocessing plants the first one of which at Kahuta, near Islamabad, went into operation in 1984. It is estimated that enough plutonium has been produced to construct between 16 and 24 bombs. The uncertainty reflects the secrecy of the program.

Since I taught in Pakistan those many years ago I have often wondered how many of my students involved themselves in these activities. The temptations must have been great. The openings in universities in Pakistan are surely very limited and many Pakistani scientists have emigrated. The ones that remained must have found opportunities in this kind of government work. Even the scientists who later chose to settle in Pakistan studied abroad whenever they could. When I was there, there was such a student about whom I had heard nothing. I doubt that very many people had heard anything about him. He was a metallurgist named Abdul Qadeer Khan. He was born in Bhophal, in what was then British India, in 1936. He emigrated with his family to Pakistan in 1952. I have not been able to find out very much about his early life. His father was headmaster of a school in Karachi. Khan did his undergraduate work in Karachi. I have not been able to learn how he was able to continue his education in West Berlin in 1961. Who sponsored him? But after a time in Germany, he moved to Holland where he got a degree in metallurgical engineering from the Technical University in Delft in 1967. This was followed by a move to Belgium where he took his PhD, again in metallurgical engineering, from the Catholic University of Leuven 1971. Like other young Pakistani degree holders he decided to work in Europe. His thesis supervisor Professor M.J. Brabers heard about a job opening at the Physical Dynamics Research Laboratory (FDO) in Almelo — a small town in Holland. Khan moved there in May of 1972. As it happened at this very time the FDO, which was a subcontractor of Ultra Centrifuge Netherland (UCN), was getting into work on the most advanced centrifuges in the world. UCN was the Dutch subsidiary of a European consortium called the Uranium Enrichment Company (URENCO). The German branch had just come up with an advance design of a centrifuge. I will explain how URENCO came into being and why it had such a design later. The people at UCN needed someone who understood the material to translate the German documents into Dutch. Khan, who knew both languages, was perfect. It gave him access to the most classified material there was on the centrifuges that could be used to enrich uranium. What no one knew at the time was that this personable young Pakistani, who had married a Dutch-South African woman, had begun in 1974 to steal this material in order to bring it to Pakistan where it became an essential component of the Pakistani atomic bomb program. In December of 1975, Khan made a hurried departure to Pakistan with blueprints for the centrifuges which he turned over to the Pakistani Atomic Energy Commission for which he shortly went to work. From that time, until he was finally unmasked in 2001, Khan created a network that sold or exchanged information and nuclear weapons technology with countries such as

Libya, Iran and North Korea. Many of his operations were run out of Dubai where interested customers could obtain a pricelist of the various offerings.The complete package, which included details as to how to make a bomb, from enriched uranium and plutonium cost about three billion dollars. It was bought by the Libyans. The Libyans may have been shortchanged because the Khan network apparently sold them a somewhat antiquated design for a bomb.

These matters are the subject of *Shopping for Bombs*, by a Security Correspondent for the BBC named Gordon Corera.[d] It is a very interesting but ultimately depressing book, depressing because of the havoc that A.Q. Kahn and his network produced in the attempt to control nuclear proliferation is still with us. Mr. Corera states at the beginning that he is not writing a biography of Khan. I wish he had included more biographical material especially about Khan's early life. Apart from Khan's being a superlative hustler, I was never able to evaluate his abilities as an engineer. How much of the bomb technology did he actually understand? I am going to begin with a brief discussion, something that Corera really does not deal with adequately, of how this rather obscure Dutch company got itself into the business of manufacturing the most advanced centrifuges in the world. This had to do with the adoption of the centrifuge technology they had gotten from Zippe.

In 1964, the Germans formed a state owned company to develop this uranium separation technology on an industrial scale. In 1970, the company was privatized and that year it signed an agreement known as the Treaty of Almelo with similar enterprises in England and Holland to create a joint company — URENCO. That is why Khan was employed by UCN to translate the plans for the Zippe centrifuge from German into Dutch.

To explain what happened next in Khan's saga I have to say a bit about the history of East and West Pakistan. When India was split in 1947, two Pakistans were created on opposite sides of India separated by a thousand miles. From the beginning there was trouble. East Pakistan had a larger population but the capital of the

---

[d]Corera, op. cit.

country — Karachi — was in the west. The people of East Pakistan, with good reason, felt economically exploited. There began a struggle for independence which was held in check by the Pakistani army until 1971, when the East Pakistani's declared their independence and the Pakistani army was defeated by the Indians. Thus Bangladesh was created. The army in Pakistan has always had a special place of honor. It has been what keeps the country from disintegrating. The spectacle of Lieutentant-General "Tiger" Nazri close to tears as he stripped his epaulettes and handed over his revolver to his Indian counterparts was so humiliating that the Pakistanis have never forgotten it. In December of 1971, Yahya Khan resigned and Zulifikar Ali Bhutto became president. He had been proclaiming the need for a Pakistani nuclear deterrence for several years, even if, as he said, people had to eat grass to achieve it. Now that he was in power this was one of his first priorities.

In January of 1972, Bhutto called a council of advisors to decide on the next step. Pakistanis from the diaspora were summoned to join their local counterparts. A colleague of mine, the one who had brokered my appointment in Pakistan, was asked to go. As far as I know he refused. At the council there were voices raised against the idea of a relatively poor third-world country spending fortunes of money on nuclear weapons when there were so many social needs. As is usually the case, they were overruled. All doubts vanished when India tested its first weapon in May of 1974. This is when Khan decided to act. He wrote a letter to Bhutto offering his services. Remarkably it got to the president and Khan was invited to present his case. This led to Khan's precipitous departure from the Netherlands taking with him the details of the Zippe centrifuge. Once Khan began to work with the official body — the Pakistan Atomic Energy Commission — he decided that it was much too cumbersome and persuaded Bhutto to set him up in his own entity which he chose to locate in Kahuta, near Islamabad. From that point on, until his final downfall, he was accountable to no one, although he did collaborate with the army. This autonomy enabled Khan to set up his network which sold or bartered nuclear technology. In his book Corera goes through the activities of the network country by country. Here I will give a few of the highlights beginning with China.

In 1945, the memo which contained a blueprint for the Los Alamos plutonium implosion bomb, which had been given to the Russians by Klaus Fuchs, was sent to Beria and became the basis of the first Russian atomic bomb. The Russians, in turn, during a brief period when their relations with the Chinese were exceptionally good, gave this blueprint to the Chinese. The Chinese in turn traded it to Khan for the centrifuge technology. It took until 1998, before the Pakistani's successfully tested their first bomb which, if Khan is to be believed, was a boosted uranium implosion weapon. The North Koreans traded missiles for the centrifuge technology. Air planes from the Pakistani air force made the deliveries and pickups showing that Khan had the cooperation of the military. The civilian government appeared to have been kept in the dark. The Iranians bought centrifuge technology and even some parts. When the IAEA inspected one of these parts they found that it contained traces of uranium that had been enriched to forty percent. One explanation was that the Iranians had a secret military program. The more likely explanation was that the part was a used piece that had been used in the Pakistani program. It was the commerce with the Libyans that finally brought the network down — if it is down.

For many years Colonel Gadafi had mixed feelings about acquiring nuclear weapons for Libya. But in 1995 he made the decision to go ahead. His representatives contacted representatives of the Khan network. Khan himself came to Tripoli several times. Gadafi decided to buy the package. This included the centrifuges, uranium hexafluoride gas to put in them for enrichment, and plans to make a nuclear weapon using the enriched uranium. In 1997, the Khan network delivered twenty so-called P-1 centrifuges of the Zippe type which enabled the Libyans to get started. But then two problems arose. The first was that the network did not have enough material on the shelf to supply the Libyans. They needed to manufacture some so they set up disguised factories in places like Malaysia. The second problem was much more serious — indeed fatal. By this time the network had been penetrated. There were moles who have never been identified. All of Khan's activities were known. One might naively think that this would have been enough to bring the network down. The problem was putting enough pressure on the Pakistani government to do this.

When the Russians entered Afghanistan we needed the Pakistanis to allow passage of the Muhajadin into Afghanistan. When the Russians left we needed the Pakistanis to help us fight the Muhajadin that we had created. To break the impasse we needed something dramatic and this was supplied by the German owned ship the BBC China. This ship had been engaged to bring nuclear material from Malaysia to Libya in October of 2003. It was tracked from the time it left port until it stopped in Italy where it was boarded and its cargo seized. This was the smoking gun which could not be ignored.

Gadafi had had second thoughts about his nuclear program and the seizing of the China was the last straw. He decided that he would use giving up his nuclear booty as a chip to trade against recognition by the United States and assurances that there would be no attempt at regime change. There and then commenced a series of negotiations which often seemed like something out of Monty Python. On December 11, as the plane with the CIA team was about the leave Tripoli, the Libyans gave them a stack of several envelopes. These, it turned out, contained the blueprints for the bomb that Khan had sold them. As far as I know these have never been released so that one does not know exactly what bomb these plans were for. If I had to make a guess it would be for the unboosted uranium fission weapon. With this now out in the open General Pervez Musharraf, now president of Pakistan, could no longer maintain the fiction that Khan's activities were a mystery to the government. On January 31 Khan was relieved of his position as special advisor to the prime minister. On February 1, Khan confessed to Musharraf what he had done and three days later he gave a speech to the nation apologizing. But for many he was still a hero — the father of the Pakistani atomic bomb. He was never tried. He was put under a comfortable house arrest in Rawalpindi where incommunicado he is at the present day. Around the world his network was rolled up, a process that is still going on. To take an example, a German named Gotthard Lerch who was accused of helping to arrange the Libyan deal went on trial in Mannheim. He had to be extradited from Switzerland. This was contested by Lerch's lawyers. But on October 6, 2008 he was convicted and sent to prison. Others connected to the network have claimed that they knew nothing of what it was actually trafficking in and as far as they were concerned the transactions were perfectly innocent. How much, if any, of the network is still operating one does not know.

In his book Corera speculates about Khan's motives. In the beginning it seems to me that they were patriotic. He seems to have lived modestly. But he clearly had a tendency towards megalomania and by the end it seems to have been a mixture of power and greed. He built a palatial mansion on the shores of Rawal Lake near Rawalpindi. In doing so he violated the local laws by spilling raw sewage into the lake. When the local authorities tried to bulldoze the house, Khan's bodyguards shot the bulldozer operator. This is presumably the house that Khan is living in today. Whatever one may think of what Klaus Fuchs did, greed was not his motive. In a fascinating book *The Man Behind the Rosenbergs*,[e] Alexander Feklisov, who was the Soviet agent who ran the Rosenbergs in the United States and Fuchs in England, describes Fuchs's adamant refusal to take money. Even being offered it Fuchs took as an insult. But on his last meeting with Feklisov on April 1, 1949, Fuchs did accept money. His older brother Gerhard was suffering from tuberculosis and was in a very expensive sanatorium in Davos. Fuchs took the money to send to his brother.

---

[e] *The Man Behind the Rosenbergs*, by Alexander Feklisov, Enigma Books, New York, 2001.

13. Who's Next?

In the spring of 1960, I made a visit to Israel at the same time that Murray Gell-Mann did. We had been working together in Paris — he on the ideas that led to the quark. He had been invited to give lectures in Israel and I was going more or less as a tourist. We ended up doing some things together. One of them was a visit to David Ben-Gurion at his home in the desert Kibbutz of Sdeh Boker. We were accompanied by Ben-Gurion's daughter Renana Leshem, a biologist, and an Israeli physicist of South African origin whose car we were driving in. The Israelis were courting Gell-Mann to come to Israel and part of the courtship was this visit. Two things stand out ineluctably in my memory. The first occurred during the visit. Ben-Gurion, who was short and somewhat wild-haired, suddenly stood up and pointed out the window to a hill. He said to Gell-Mann, "Do you see those trees up there?" We both looked at the bare hill and Gell-Mann commented that there were no trees. Ben Gurion responded by saying that in then years there would be trees. The other incident occurred on the drive back to Tel Aviv. It was dark and somewhere in the desert our driver took the wrong turn. This became evident when we came to what was clearly a fenced gatepost with armed guards. In the distance I could see a large structure that was brilliantly illuminated. We explained to the guards our navigation error and as we were about to leave I saw a sign that said "Dimona." We had stumbled upon the Israeli nuclear weapons facility.

If we had actually visited the site we would have come upon several hundred people speaking French.[a] To understand this we must give a brief account of how this program got started. As soon as Israel had won the "War for Independence" in 1948 Ben Gurion began thinking of a nuclear deterrent. After the experience of the Holocaust he had no faith in security guarantees by external powers. He imagined a scenario where there might be a concerted attack by Israel's neighbors which would destroy much of the country before anyone could or would provide help. When Ben Gurion had this vision of a nuclear Israel there were very few Israeli nuclear physicists and no real infrastructure for this kind of thing, to say nothing of money. But Ben Gurion was not an individual who was easily discouraged. He had as allies the chemist Ernst David Bergmann who organized the scientific effort and above all Shimon Peres the *éminence grise* who was responsible for all the practical negotiations. It was Peres who ultimately made the arrangements with the French. Before this could happen Israel had to assemble a cadre of nuclear scientists. This it did by sending some of their young physicists abroad where they could study in European and American institutions. I knew a few of them. For example Harry Lipkin, who was born in the United States and emigrated to Israel in 1950 was sent to spend a year at Saclay in France studying reactor physics in 1953 as was Amos de Shalit who had studied at MIT. Both of these men were outstanding physicists. By the way, our driver that evening had gotten his PhD at Harvard about the same time I got mine. By the late 1950s the Israelis had put together a group of nuclear physicists about as good as any. The question that any such group interested in building an

---

[a]An excellent account of this history can be found in *Israel and the Bomb* by Avner Cohen, Columbia University Press, New York, 1998.

atomic bomb from scratch must grapple with from the beginning is what kind of bomb to build.

There are really two choices. One can try to build a bomb whose explosive fissile material is uranium or one can try to build one whose explosive fissile material is plutonium. Either choice presents technical challenges. If uranium is chosen then one has to deal with "enrichment." There are a variety of techniques for doing this of which the most popular presently is the centrifuge. The Israelis have a centrifuge program whose details are secret, but they chose to go the plutonium route which requires building a nuclear reactor. Such a reactor also requires uranium — something like a hundred tons to keep it running at any given time. Over its lifetime it uses thousands of tons. The only uranium Israel had was in the phosphates in the Dead Sea. They invented a process to extract this uranium. They eventually sold this process to the French. But this Dead Sea phosphate could not produce anything like the amount they needed. They would have to buy the rest. But that was not the only thing they had to buy.

Most reactors require that natural uranium that it ultimately used for fuel be enriched to something like three and a half percent. However, the Israelis chose to build a type of reactor for which the uranium required no enrichment. This has to do with the "moderator."

The fissile isotope uranium-235 can be fissioned by neutrons of any energy while uranium-238 requires neutrons above a threshold energy. This is the distinction between fissile nuclei such as uranium-235 and any of the isotopes of plutonium, and fissionable isotopes such as uranium-238. When a nucleus is fissioned it is split into lighter nuclei. Energetic neutrons are also emitted so a chain reaction is possible. These neutrons move with speeds about the tenth the speed of light. This is why a chain reaction can set off an explosion. A kilogram of uranium-235 can be completely fissioned and the energy released in about a microsecond. An explosion is the creation of large amounts of energy in a small space in very little time. But as we have already noted there is a paradoxical fact about fission that can only be explained using quantum mechanics. The less energy the incident neutron has the more probable it is that it will fission a uranium-235 nucleus. One does not want to slow down the chain reacting neutrons in a bomb to take advantage of this because one wants the fission energy to be produced as rapidly as possible. But in a reactor one does want to slow the neutrons down — hence the moderator.

The rapidly moving neutrons produced in fission collide with the nuclei of the moderator. This transfers some of the energy from the neutron to a moderator nucleus. The best moderator in principle would be ordinary hydrogen whose nucleus is a single proton. The proton and the neutron have sensibly the same mass so that in a collision much momentum is transferred. Think of two billiard balls colliding. But protons can capture neutrons forming a nucleus of "heavy hydrogen" — the deuteron. This removes the neutron from the fission cycle which is undesirable. The next best nucleus for these purposes is the deuteron. It also captures neutrons but very little. To exploit the deuteron requires "heavy water" — water in which the protons in the water molecules are replaced by deuterons. In ordinary water about one in sixty-four hundred molecules contain the deuteron as opposed to a proton. The problem is to separate out the heavier molecules. There are various ways of doing this which make use of the fact that because of the mass difference there are differences in reaction rates. For example, electrolysis — which uses electric currents — is employed to first separate the hydrogen — heavy and light — in water from the oxygen. Ordinary hydrogen separates more rapidly which leaves the heavy hydrogen which can then be combined with oxygen to make heavy water. The Israelis invented a process for doing this but nonetheless they needed to buy most of the supply they needed, at least ten tons. If they could find this amount then they could use natural uranium in their putative reactor which would have two advantages. On the one hand they would not have to enrich the uranium used as a fuel and on the other it would maximize the plutonium production. This is because plutonium is created in a reactor in a series of reactions that begin with the absorption of a neutron by a uranium-238 nucleus. So the greater the ratio of uranium-238 to uranium-235, the greater the production. One always needs some uranium-235 for fission. The question the Israelis had to deal with was where were they going to get the heavy water?

At first they tried to buy it from the United States but it was made clear that they would only be sold technology appropriate to a small research reactor which might have a plutonium production capacity of the order of grams per year. Kilograms are needed in a nuclear weapon. They next turned to Norway. Why Norway? In 1934 at Vermorsk the Norwegian company Norsk Hydro built the first commercial heavy water production facility in the world. It used electrolysis and could produce about twelve tons of heavy water a year. When during the war the Germans — Heisenberg and company — began designing nuclear reactors they too wanted to use heavy water and natural uranium. They proposed to get this from Norway which they had occupied. The British got wind of this and with the aid of the Norwegian resistance they managed in some daring raids to sabotage the plant. In February 1944, a ferry boat — the *Hydro* — which was thought to be carrying drums of heavy water to Germany — was sunk crossing Lake Tinsjo. Many years later the drums of the alleged heavy water were retrieved and it turned out that the percentage of heavy water in them was too small to have been used in a reactor. One wonders if the Norwegians had some of this history in mind in 1959 when, after some serious negotiations, they sold the Israelis twenty tons of the stuff. They had extracted an agreement that it would only be used for "research." The only kind of research the Israelis were interested in was the manufacture of plutonium which they did not tell the Norwegians. This left the matter of the reactor itself. Enter the French.

It became clear to Ben Gurion that buying this technology from the French or anyone else was going to be expensive. Therefore in the 1950s he organized a fund raising campaign just for this purpose. It is not clear what he told the potential donors. The Israelis have never admitted that they have a nuclear weapons program. But whatever he told them he managed to raise about forty million dollars. That French companies got a good deal of this can be traced to the miscalculation of the Egyptian strongman Gamal Abdul Nasser and the adroit diplomatic skills of people like Shimon Peres. In 1956 Nasser closed the Suez Canal to international traffic. Peres had already had negotiations with the French on nuclear cooperation but these had stalled. He was now approached about a joint Israeli, French operation to reopen the canal. The Israelis agreed provided that the French would sell them technology for their reactor. This was not stated explicitly in the agreement but it was understood by both parties. When the Russians threatened nuclear force if the Israelis did not withdraw, Peres made the case with the French that they, like the French needed their own nuclear deterrence. Hence the agreement to aid in the construction was reached. But in 1958 General de Gaulle became prime minister and he decided to end the arrangement. Ben Gurion met with de Gaulle and Peres worked out a compromise. The French companies who were already involved would be allowed to continue but that no new cooperation would be allowed. That is why if we had visited Dimona in 1960 we would have heard French. The reactor was under construction. But what would the energy dimensions of the nascent reactor be?

Two units of power output for reactors are generally used — megawatts thermal and megawatts electric. Megawatts thermal is a measure of the raw power that the reactor can generate while megawatts electric is a measure of the electric power that can be produced. A large reactor might generate three thousand megawatts thermal but only one thousand megawatts electric. The Israelis had no interest in using their reactor to generate electricity so the only relevant measure is the megawatts thermal. The French had signed on to build a reactor like the one they had in Marcoule which generated about forty megawatts thermal. They noticed that their Israeli counterparts were redesigning things like the cooling systems so that the reactor could be upgraded. Indeed, because of secrecy no one not on the inside knows what was actually built and what has been upgraded. There are a variety of estimates. These influence estimates of the amount of plutonium that has been produced which in turn influences the estimates of how many nuclear weapons the Israelis actually have. Some estimates give one hundred and some give four hundred. This is an ambiguity that the Israelis have been determined to keep.

To give another example, the French Saint Gobain chemical company was brought in to build a facility for separating the plutonium that had been created in the uranium fuel rods, from the fuel rods. This is a non-trivial matter. The French had supplied all of the uranium used initially and some sort of understanding had been reached that the irradiated fuel rods would be sent back to France. The Israelis kept referring to this when they had difficulty with people like President Kennedy. In fact it seems that one shipment of fuel rods was sent to France and the small amount of plutonium returned to Israel. As soon as the Israelis got the reprocessing plant to work they did their own reprocessing and of course kept the plutonium. Ben Gurion and Peres during the beginning of the program kept all these arrangements to themselves. There was never a national debate as to whether or not to have nuclear weapons program. The existence of Dimona was not revealed to the Israeli people until December of 1960 in a speech by Ben Gurion in which he insisted that the reactor was being built for purely scientific purposes. On other occasions he claimed that power from the reactor would be used to desalinate sea water. In fact there was never any purpose in building the reactor except to make plutonium for nuclear weapons. All of this deception might have continued indefinitely if it had not been for Mordechai Vanunu.

Vanunu was born in Morocco in 1934. At age nine emigrated with some of his brothers and sisters to Israel where he attended ultra-orthodox schools. When he became of age he served in the army. In 1976 he began working at Dimona as a night shift manager. What he was told about the purpose of the facility is not clear. He apparently did not need to know any physics to do his job. During the day he studied philosophy at the Ben Gurion University in the Negev. He graduated in 1985. During his period at Dimona his politics became more and more radical Left and in 1985 he was fired. He did what many people at loose ends did, he began travelling to places like Nepal. He thought for awhile becoming a convert to Buddhism but in 1986 he landed in Sydney, Australia where he supported himself with various odd jobs such as dishwashing and driving a taxi. He began going to a local church and converted to Christianity and was baptized. While in Sydney he met Peter Hounam a journalist from *The Sunday Times* in London. He told Hounam about Dimona and had photographs to back up his story that he had illegally taken while at work. Vanunu's revelations were published in the *Times* on the fifth of October. The Israelis were determined to arrest him and a Mossad agent named Cheryl Bentov persuaded him to go with her to Rome for a "holiday" where the Mossad captured him and took him to Israel. He was tried and sentenced to eighteen years much of which he spent in solitary confinement. Since his release he has been in constant trouble with the authorities and claims that he has been persecuted since he is a Christian. What did Vanunu reveal? The details about the reactor itself were the least of it.

Not far from the reactor there is a two story building that looks like any other building. The portion that a casual visitor would see consists of offices, storage facilities and a worker's canteen. But what such a visitor would not see are the elevators that descend to the various levels where the actual work is being done. There are levels for example for uranium enrichment and plutonium separation. To me the most interesting level is the one that produces lithium-deuteride. If you are not familiar with nuclear weapons technology this will probably not mean anything. If you are, what it means is that any program that uses it has gone beyond the rudimentary nuclear weapons stage. Why?

Lithium-deuteride is a grayish stable solid compound of lithium and deuterium — the deuteron again. If the lithium is subjected to bombardment by neutrons it undergoes a reaction in which helium, and more importantly, super heavy hydrogen-tritium is produced. The tritium nucleus has two neutrons and a proton. It is also unstable with a half-life of about twelve and half years. When a mixture of tritium and deuterium is subjected to the kinds of temperatures and pressures that a nuclear device can produce they fuse and very energetic neutrons are produced. These neutrons can cause further fissions of whatever plutonium is still left unfissioned after the first explosion. This boosts the yield of the bomb and such weapons are known as "boosted" fission weapons. This reaction is also used in hydrogen bombs. Hence when a country with a nuclear weapons program has a facility for making lithium deuteride one knows that they are in the business of making advanced nuclear weapons. The Iranians have a facility for producing heavy water and hence deuterium. Whether they are trying to produce lithium deuteride I do not know.

We are all too familiar with the rantings of Mahmoud Ahmadinejad. The assumption that people made who favored negotiation with Iran was that he was a figure head and that the real power lies in the hands of the clerics who might be more reasonable. I have my doubts when it comes to the nuclear issue, I cite again as an example Akbar Hashemi Rafsanjani who was president of Iran from 1989 to 1987. It is known that he was hostile to Ahmandinejad and some of the ideologies of the Islamic Republic. This may be true but on the matter of nuclear weapons he has said things that are much more worrisome than anything that Ahmandinejad had said. Here is something he said at Friday prayers in December of 2001. "If a day comes when the world of Islam is duly equipped with the arms Israel has in its possession, the strategy of colonization would face a stalemate because application of an atomic bomb would not leave anything in Israel but the same thing would just produce damages in the Muslim world." The latter part of this statement persuades me that Rafsanjani has no comprehension of what a nuclear weapon is. Part of this may be an unintended consequence of the ban against testing above ground. The last such test was by the Chinese in 1980. The number of people who have actually seen a nuclear explosion grows fewer and fewer. I saw two in Nevada in August of 1957 and I will never be able to forget the experience. Over the years I have tried to figure out a way to dramatize what these explosion are. I finally hit on the Ryder truck. At 9.02 am on April 19, 1995 Timothy McVeigh exploded a Ryder truck packed with explosives in front of the Alfred P. Murrah Federal Building in Oklahoma City. It destroyed half the building. One hundred sixty eight people were killed and four hundred and fifty were wounded. The truck contained five thousand pounds — two and half tons — of high explosive. The Nagasaki bomb that destroyed the entire city produced twenty **thousand** equivalent tons of TNT-8000 Ryder trucks — to say nothing of the radiation. A hydrogen bomb is equivalent to millions of Ryder trucks. That is what we are talking about. An attack on Israel would produce a response with hundreds of planes and missiles all nuclear armed. Some would not get through, but some would. A very small number of atomic bombs would be enough to reduce Iran to rubble. Rafsanjani is talking about national suicide. How do you deter someone or some country that is willing to commit suicide?

Many years ago I met Benjamin Netanyhu at a party. I was introduced as a nuclear physicist who had been at Los Alamos. It is true that I had spent a summer at Los Alamos and had written a couple of unimportant papers on nuclear physics but in truth I knew nothing about nuclear weapons. But for some reason we began talking about them. I remember Netanyhu saying to me in a very serious tone that if Israel ever had its back to the wall it would use them. I think that it is clear that nothing has changed his mind.

14. A Memorandum that Changed the World

Reprinted with permission *American Journal of Physics* **79**, 440, 2011

In the summer of 1939 the Austrian born physicist Otto Frisch, who was in Copenhagen, received an invitation from Mark Oliphant, a professor of physics at Birmingham, to come to Birmingham for the summer to see if some arrangements might be made for Frisch to emigrate to Birmingham. War came that fall and Frisch could not return safely to Denmark. He was given a lectureship that enabled him to stay.

The previous Christmas he had gone to visit his aunt Lise Meitner in Kungalv in Sweden. She had just received a letter from Otto Hahn, the radiochemist in Berlin with whom she had collaborated for many years before she was forced to flee Germany. Meitner and Hahn had been using neutron sources to bombard various elements including uranium. Hahn had continued the experiments with Fritz Strassmann, another chemist. They had found something they did not understand. Instead of finding elements in the residual bombarded uranium that had masses comparable to uranium, they found barium, which was somewhere in the middle of the Periodic Table. They were stuck for an explanation and Hahn appealed to Meitner. She and Frisch went for an excursion in the snowy woods — he on skis and she on foot. During that outing they realized that by using the liquid drop model of the heavy nuclei, what Hahn and Strassmann had done was to fission the uranium nucleus. One product was barium and the other was krypton with possible neutrons in addition. They worked out the energies and realized that about 200 MeV would be released, a very large amount of energy compared to any chemical reaction.

Frisch returned to Copenhagen after this visit with his aunt (his aunt remained in Sweden) and told Bohr about what he and Meitner had concluded. Bohr's reaction was like a great many other physicists when they heard about it — it was so obvious why they had not thought of it. As it happened Bohr was about to leave for the United States where he was going to spend time at the Institute for Advanced Study which was then located at Princeton University. He was accompanied by his assistant Lèon Rosenfeld. On the Atlantic crossing Bohr had a blackboard setup in his cabin, and he and Rosenfeld went over the theory. He had promised Frisch and Meitner not to say anything until their paper appeared, but he had forgotten to tell Rosenfeld who went at once to Princeton and told everybody.

Even in those pre-internet days the news spread all over the country and abroad and fission experiments proliferated. Among the first were those of Frisch, who was primarily an experimentalist. He looked for ionization pulses, which were produced by the charged fission fragments. Others found the additional neutrons — on average a little over two — produced in the fission process. Ironically, the barium-krypton split which Hahn and Strassmann found was a relatively unlikely nuclear splitting.

In his autobiography, *What Little I Remember*,[1] Frisch writes "In all this excitement we had missed the most important point. It was Christian Møller, a Danish colleague, who first suggested to me that the fission fragments the two freshly formed nuclei might contain enough surplus energy to each eject a neutron or two; each of these might cause another fission and generate more neutrons. By such a "chain reaction" the neutrons would multiply in uranium like rabbits in a meadow! My immediate answer was that in that case no uranium ore deposits could exist; they would have blown up long ago by the explosive multiplication of neutrons in them. However, I quickly saw that my argument was too naive; ores contained lots of other elements which might swallow up the neutrons; and the seams were perhaps thin, and then most of the neutrons would escape. So from Møller's remark the exciting vision arose that by assembling enough pure uranium with appropriate care, one might start a controlled chain reaction and liberate nuclear energy on a scale that really mattered. Many others had the same thought, as I soon found out. Of course the specter of a bomb — an uncontrolled chain reaction — was there as well; however, for awhile anyhow, it looked as though it need not frighten us. That complacency was based on an argument of Bohr, which was subtle but appeared quite sound."[2]

The February 15, 1939 issue of the *Physical Review* contained a two page article by Bohr that changed everything.[3] It had the unremarkable title "Resonance in uranium and thorium disintegrations and the phenomenon of nuclear fission." The essential point came in the penultimate paragraph and might easily have been overlooked. Bohr noted that what counts in the fission process is the formation of a compound nucleus after the incident neutron has been absorbed. For ^{238}U, the common isotope, it is ^{239}U, while for ^{235}U, the rare isotope, it is ^{236}U. The latter is an even-even nucleus, while the former is an even-odd nucleus and hence more loosely bound. This difference in binding energies results in a difference in the mass defect between the initial state of the neutron and one of the isotopes of uranium and the compound nucleus. This difference for the uranium isotopes is about 1.7 MeV. The extra energy due to this mass defect goes into the excitation energy of the compound nucleus and is what is responsible for its fission. To make ^{238}U fission we must supply an energy of at least 1 MeV from the incident neutron while for ^{235}U this energy is supplied by the mass defect and hence neutrons of any energy can cause fission. This energy threshold or lack of same is the difference between a "fissionable" and a "fissile" nucleus.

Frisch's complacency and Bohr's as well had to do with the realization that the dominant isotope could not make a self-sustaining chain reaction. Much of the spectrum of the emitted neutrons would be below the threshold energy for fission. Thus, to make such a chain reaction would require the separation of isotopes on an industrial scale. Bohr ruled this separation out because he said it would take the resources of an entire country. Actually, it took the resources of three: Great Britain, the United States, and Canada. However, up to this point no one had actually determined how much ^{235}U was needed to make a "critical mass" — a mass above which the chain reaction would be self-sustaining. Enter Rudolf Peierls.

Peierls was born in Berlin in 1907 which made him three years younger than Frisch who was born in Vienna. Both men were of Jewish ancestry. Peierls, who was a theorist, took his degree in Munich from Arnold Sommerfeld. Sommerfeld was one of the great teachers of physics with such star pupils as Heisenberg, Pauli, and Bethe. As Peierls moved up the academic ladder, he too set up schools of theoretical physics, first at Manchester, then Birmingham, and finally in Oxford. For a while he was an assistant to Pauli who complained that Peierls was so fast that after telling you about an idea he would tell what was wrong with it before you had a chance to grasp the original idea. He made significant contributions in every branch of theoretical physics. When Hitler came to power, Peierls was in Cambridge on a Rockefeller Scholarship. He stayed in Britain in different positions until he was named a professor of physics in Birmingham in 1937.

When fission was discovered, it was natural that Peierls would take an interest. The first question he addressed was how to make a chain reaction. In 1939 he published a paper in the *Proceedings of the Cambridge Philosophical Society*[4] entitled "Critical conditions in neutron multiplication." The first paragraphs of this paper read:

"It is well known that a single neutron may cause a nuclear reaction chain of considerable magnitude, if it moves in a medium in which the number of secondary neutrons which are produced by neutron impact is, on the average, greater than the number of absorbed neutrons. From recent experiments it would appear as if this condition might be satisfied in the case of uranium.

Such multiplication of neutrons can only take place if the path traveled by each neutron in the body is long enough to give it a sufficiently high chance of making a collision. It seems of some interest to discuss the dependence of the phenomenon on the size of the body."[5]

Let me restate these ideas in somewhat different language. Suppose a solid sphere of uranium has been assembled. In his paper, Peierls does not specify the isotope. In the interior of the sphere neutrons are being produced in the fission products. Within the sphere the neutrons can collide elastically and inelastically, they can be absorbed, or they can produce more fission products. However, the neutrons can also escape through the surface of the sphere. What we want to know is the critical radius of the sphere — the size at which the number of neutrons that escape just balances the number of neutrons that are created. If we know this critical radius, we know the volume of the sphere and from the mass density we then know the critical mass. At just this mass there is no self-sustaining chain reaction produced. We need a "supercritical" mass. This idea was tested and confirmed in the Godiva experiments at Los Alamos.[6] Frisch participated and gave its name because bare masses were being used. In an actual bomb the fissile sphere is surrounded by a heavy metal casing that reflects neutrons and thus enhances the fission reactions.

The fission mean free path for the neutrons, the average distance between fissions, is roughly the order of magnitude of the critical radius. The Peierls theory refines this estimate. The mean free path for fission is by definition $r_f = 1/n\,S_f$. Here, n is the number density of the uranium nuclei in the sphere and S_f is the fission cross section. Curiously, Peierls did not put in any numbers to estimate this mass. This estimation was left to Frisch. Frisch had a good idea on the order of magnitude of n, but not much of an idea about S_f. In his paper, Bohr wrote that this cross section for fission can never "exceed nuclear dimensions."[7] If we apply this statement literally and take the radius of a uranium nucleus to be about 7 F $(10^{-15}\,\mathrm{m})$, equal to about 7×10^{-13} cm, then the area is about $1.5 \times 10^{-24}\,\mathrm{cm}^2 = 1.5\,\mathrm{b}$.

What Frisch and Peierls did not know was that this cross section was being measured for slow neutrons on trace amounts of $^{235}\mathrm{U}$ that had been electromagnetically separated in a collaboration between Alfred O. Nier of the University of Minnesota and E.T. Booth, J.R. Dunning, and A.V. Grosse of Columbia University. In March of 1940 they announced results of somewhere between 400 and 500 b.[8] The huge discrepancy between this result and Bohr's estimate has to do with quantum mechanics. Once the neutron is slowed down to where its de Broglie wavelength is about the size of the target — the uranium atom — the classical geometric picture no longer applies. A thermal neutron which has an energy of about 0.025 eV has a de

Broglie wavelength of about 1.8 Å. All Frisch could do was to make a guess at the fission cross section, and he took it to be about 10 b or 10^{-23} cm^2, which for the energy region that is relevant — say 2 MeV — is one order of magnitude too large. This overestimate reflects itself in the critical mass which goes as the cube of the radius and hence of the inverse cross section. Frisch found a critical mass that was the order of a pound or so — much too small. The actual critical mass of ^{235}U is about 115 lbs.

Having found this small critical mass Frisch alerted Peierls and began thinking of ways in which he could actually separate the uranium isotopes. First the latter.

Frisch was familiar with a method that had been invented in 1938 in Germany by the German physical chemist Klaus Clusius and his younger colleague Gerhard Dickel.[9] In essence the separation apparatus consists of a vertical tube with a wire down the middle that can be heated. If a gas with different isotopes is introduced, the lighter isotope concentrates at the heated element and accumulates at the top, while the heavy element, as a kind of countercurrent, accumulates at the bottom. In 1939 Clusius announced the separation of chlorine isotopes using his method and he began collaboration with Paul Harteck and others to begin trying to separate uranium isotopes in a uranium hexafluoride gas. Frisch had no way of knowing that this method was unsuccessful. Apart from the corrosive effects of this gas, the high temperatures involved created instability in the uranium hexaflouride molecules, and the Germans switched their attention to using centrifuges. However, Frisch mastered enough of the difficult theory to realize that the efficiency of the process could be improved if a bigger tube rather a thin wire was used. He needed a glass blower to make the tube and these people had radar as a priority. While he was waiting for the equipment, he received an invitation to write a report for the British Chemical Society on advances in nuclear physics, and he included a section on fission and its prospects.

Neither Frisch nor Peierls was British citizen. In fact, they were classified as enemy aliens, which meant that they could not work on any secret military program including radar. However, Oliphant, Peierls' colleague at Birmingham, got around this problem by posing questions to Peierls that were in the guise of abstract problems in electromagnetism. Peierls knew they were connected to radar and Oliphant knew that he knew, but the security fiction was preserved.

Peierls now took the prospect of nuclear weapons very seriously, which led to two memoranda, one of which I am going to deconstruct. After I finish this exercise I will discuss what happened to these memoranda. I will present the memorandum line by line and add my comment comments in square brackets.

The memorandum was titled "On the construction of a 'super-bomb' based on a nuclear chain reaction in uranium" March 1940.[10] The date is significant because it sets limits of what Frisch and Peierls knew. For example, the paper by Nier *et al.*[8] was published after this date. In the following all quotations are from the memorandum, and my comments are contained within brackets.

"The possible construction of superbombs based on a nuclear chain reaction has been discussed a great deal and experiments have been brought forward which seemed to exclude this possibility. We wish here to point out and discuss a possibility that seems to have been overlooked in these earlier discussions." [I wonder what discussions are being referred to. The new possibility is the role of ^{235}U.]

"Uranium consists essentially of two isotopes, ^{238}U 99.3% and ^{235}U 0.7%. If a uranium nucleus is hit by a neutron, three processes are possible: (1) scattering, whereby the neutron changes directions and if its energy is above 0.1 MeV, loses energy; (2) capture, when the neutron is taken up by the nucleus; and (3) fission, i.e., the nucleus breaks up into two nuclei of comparable size, with the liberation of an energy of about 200 MeV." [I confess that when I first read the memorandum, I found the first of the three possibilities incomprehensible as discussed by these authors, but I have been able to deconstruct what they meant.[11] For elastic scattering the incident neutron can lose energy. If we average over all angles and call the average final energy $E_{f,av}$ and the initial energy E, then $E_{f,av}/E = 91 + (A - 1)^2)/(A + 1)^2/2$. Here, A is the mass number. This expression tells us why heavy elements such as uranium are poor moderators. If we substitute A = 235 in this expression, we find $E_{f,av}/E = 0.99$, which means that it takes a couple of thousand elastic collisions to thermalize neutrons with a uranium moderator. The separation of the energy levels in these heavy elements near the ground state is about 0.1 MeV. This energy is the threshold for inelastic scattering and is a measure of the energy loss in such an event.]

"The possibility of a chain reaction is given by the fact that neutrons are emitted in the fission and that the number of these neutrons per fission is greater than 1. The most probable value for this figure seems to be 2.3, from two independent determinations.

"However it has been shown that even in a large block of ordinary uranium no chain reaction would take place since too many neutrons would be slowed down by inelastic scattering into the energy region where they are strongly absorbed by ^{238}U.

"Several people have tried to make chain reaction possible by mixing uranium with water, which reduces the energy of the neutrons still further and thereby increases their efficiency again. It seems fairly certain, however, that even then it is impossible to sustain a chain reaction." [I wonder who these people are. If a heavy water moderator is used, then chain reactions can be sustained with natural uranium, which is what the Germans tried to do in their reactor program.]

"In any case, no arrangement containing hydrogen and based on the notion of slow neutrons could act as an effective superbomb because the reactions would be too slow. The time required to slow down a neutron is about 10^{-5} s, and the average time lost before a neutron hits a uranium nucleus is even 10^{-4} s." [Unfortunately Frisch and Peierls do not give us any information on how they arrived at these numbers. I will try to make their argument and use the data that is now available. The idea of an arrangement using hydrogen was also later considered at Los Alamos. The motivation was to take advantage of the 1/v law for fission cross sections. We have seen that Nier *et al.*[8] measured these cross sections for slow neutrons to be several hundred barns. This method was also abandoned at Los Alamos because it was too slow. Let us try to understand the 10^{-5} s time required for the thermalization of the neutrons; that is, the reduction of the average neutron energy from about 2 MeV to the thermal energy of 0.025 eV by elastic collisions of neutrons with hydrogen. For the sake of argument I will take the elastic cross section to be $20\,\text{b} = 2 \times 10^{-23}\,\text{cm}^2$. We will use this number to estimate the mean free path for elastic scattering. I will take the number density of water to be $3.3 \times 10^2 \, \text{cm}^3$. Therefore, the mean free path for elastic scattering is about 1.5 cm. It takes about 18 elastic collisions of neutrons with water molecules to thermalize the neutron. During this time it travels about 27 cm, which means that to have a thermalization time of 10^{-5} s, the speed of the neutrons would have to be about 3×10^6 cm/s, which is somewhat faster than the thermal speed of 2.2×10^5 cm/s but considerably slower than the fission neutrons which move at a speed of about a tenth that of light. I have no way of knowing if this reasoning is what Frisch and Peierls did but their answer seems reasonable. I will not derive the mean free path for fission here but leave it to later. With their choice of parameters Frisch and Peierls claim that the mean free path is 2.6 cm. If we divide this value by the thermal speed, we obtain $\sim 10^{-4}$ s as noted previously. Later I will explain why this time is much too short for the slow neutrons to play a role in the explosive chain reaction.]

"In the reaction, the number of neutrons should increase exponentially, like e^t where t would be at least 10^{-4} s." [Later I am going to use this exponential to determine the time it takes to fission a kilogram of ^{235}U using the value of f for fast neutrons. Often in this discussion, assuming that two neutrons are created per fission, this exponential is replaced by 2^x where x is the number of generations. This doubling is not really correct and should be replaced by an exponential tail. I also emphasize that most of the neutrons are created in the last couple of iterations.] "When the temperature reaches several thousand degrees the container of the bomb will break and within 10^{-4} s the uranium would have expanded sufficiently to let neutrons escape and therefore stop the reaction. The energy liberated would, therefore, be only a few times the energy required to break the container, i.e., of the same order of magnitude as with ordinary high explosives." [What is being said here is that with an exponential folding time of 10^{-4} s for the creation of neutrons, the uranium will have expanded enough so that the density is once again below critical and the bomb shuts off before producing a substantial amount of energy.]

"Bohr has put forward strong arguments for the suggestion that the fission observed with slow neutrons is to be ascribed to the rare isotope ^{235}U and that this isotope has on the whole, a much greater fission probability than the common isotope ^{238}U. Effective methods for the separation of isotopes have been developed recently, of which the method of thermal diffusion is simple enough to permit separation of a fairly large scale." [They are still optimistic about the Clusius–Dickel method which did not work out.] "This permits, in principle, the use of nearly pure ^{235}U in such a bomb, a possibility which apparently has not so far been seriously considered. We have discussed this possibility and have come to the conclusion that a moderate amount of ^{235}U would indeed constitute an extremely efficient explosive.

"The behavior of ^{235}U under bombardment with fast neutrons is not known experimentally, but from rather simple theoretical arguments it can be concluded that almost every collision produces fission and that neutrons of any energy are effective." [I am a little puzzled by the statement that almost every collision produces fission. The cross section for inelastic scattering, for example, is comparable to that of fission. Frisch and Peierls overestimated the size of the fission cross section. There are about five elastic collisions between fissions.] "Therefore it is not necessary to add hydrogen, and the reaction, depending on the action of fast neutrons, develops with great rapidity so that a considerable part of the total energy is liberated before the reaction gets stopped on account of the expansion of the material.

"The critical radius r_0, i.e., the radius of a sphere in which the surplus of neutrons created by the fission is just equal to the loss of neutrons by escape through the surface, is for a material with a given composition in a fixed ratio to the mean free path of the neutrons and this in turn is inversely proportional to the density. It therefore pays to bring the material to the densest possible form, i.e., the metallic state probably sintered or hammered." [The mass goes as ρ times the volume of the sphere, that is, ρr^3. Because the mean free path goes as $1/r$, the critical mass goes as $1/r^2$. This quadratic dependence has a very important application in implosion weapons. Before the sphere is imploded the mass is subcritical at normal densities. However, when the sphere is shrunk the density increases and the same mass becomes supercritical.] "If we assume for ^{235}U, no appreciable scattering, and 2.3 neutrons emitted per fission, then the critical radius is found to be 0.8 times the mean free path. In the metallic state (density 15) and assuming a fission cross section of 10^{-23} cm^2, the mean free path would be 2.6 cm and r_0 would be 2.1 cm corresponding to a mass of 600 g. A sphere of metallic ^{235}U of a radius greater than r_0 would be explosive and one might think about 1 kg as a suitable size for a bomb."

[This paragraph is arguably the most significant passage in the memorandum and it is substantially wrong. The Hiroshima bomb required 64 kg of uranium, 50 kg of which were 89% enriched and the remaining 14 kg were 50% enriched, leading to a total of about 52 kg of ^{235}U. As I have mentioned, we can only wonder if at this time the British would have pursued their program with the same intensity if the real figures had been known. Now to the production of these figures.

I will begin by deriving the correct mean free path for fission by fast neutrons of ^{235}U. First we need the cross section. Then I will discuss the sort of numbers that were available to Frisch and Peierls at the time they wrote their memorandum. A modern value is $S_f = 1.24$ b with a small error. We next need the number of ^{235}U nuclei per cubic centimeter for metallic uranium. The density of metallic ^{235}U is about 19 g/cm^3. Frisch and Peierls used 15 g/cm^3. Each ^{235}U nucleus has a mass of about 3.9×10^{-22} g. Hence, the number per cubic centimeter is about 4.9×10^{22}. Thus, the mean free path is about 16.5 cm. With their various assumptions Frisch and Peierls found 2.6 cm.

They claim that the critical radius is 0.8 times the mean free path. Using their various numbers we have, by multiplying the volume V times the density, Mc $= 4.3 \times 0.8 \times 2.6^3$ cm$^3 \times 15$ g/cm$^3 = 565$ g. The volume turns out to be 38 cm^3. For comparison the volume of a tennis ball is about 137 cm^3. It is little wonder that Peierls and Frisch were alarmed.

Let me redo the numbers using an expression for the critical mass that can be derived reasonably straightforwardly. This methodology is less sophisticated than the 1939 paper of Peierls.[4] Hence, we do not expect it to yield a precise answer. The expression for the critical radius in terms of the mean free path is $r_c = \pi/3(r_f/(v - 1))$, where v is the average number of neutrons produced per fission, which I will take as 2.5. Here r_f is the fission mean free path which is about 16.5 cm. This expression gives a critical radius of 24.4 cm, and with the correct density we obtain a critical mass of about a metric ton. This value is much too big and shows that this calculation must be done with care. Indeed this expression is only applicable in the approximation used by Frisch and Peierls that there is no elastic scattering. Otherwise we have to replace r_f by $(r_f r_{total})^{1/2}$ where r_{total} is the total mean free path including elastic scattering. Because $(r_f r_{total})^{1/2}$ is less than r_f the fission mean free path, the critical mass is less.

Now I turn to the numbers used by Frisch and Peierls. This consideration is conjectural because there are no references or acknowledgments in their memorandum. The date of their memorandum is March 1940, which presumably means that anything published after that date would be inaccessible to them unless it had been communicated by preprint or private communication.

They did know of Bohr's February publication of the role of ^{235}U.[3] However, we can look for clues in the papers that they had published prior to the memorandum, starting with the 1939 paper by Peierls.[4] There is only one relevant reference in this paper, and it is by Perrin.[13] Perrin apparently had not heard of Bohr's work so he seems to have tried to make a chain reaction using natural uranium. He somehow arrived at a critical radius of 130 cm and a critical mass of 40 metric tons. He concluded that it is impossible to make a chain reaction using fast neutrons and natural uranium and suggests slowing them with hydrogen. The only relevant subject in Perrin's paper is the idea of critical mass.

A paper that Frisch and Peierls must have seen is entitled "The scattering by uranium nuclei of fast neutrons and possible neutron emission resulting from fission" by Goldstein *et al.*, published on July 29, 1939.[14] This paper presents a measurement of the fission cross section of fast neutrons incident on uranium. There are four problems with this measurement: the uranium is not separated, the uranium is in the form of uranium oxide, the neutron energy spectrum is not precisely known, and most importantly, the measurement measures the total cross section, which includes elastic and inelastic scatterings as well as fission. Nonetheless, the authors conclude that the fission cross section is about 10 b. They note the agreement of this number with an earlier measurement done at Columbia University by a group that included Fermi. It is therefore not surprising that Frisch and Peierls took 10 b as the fission cross section when they estimated the critical mass.]

"The speed of the reaction is easy to estimate. The neutrons emitted in the fission have velocities of about 10^9 cm s^{-1} and they have to travel 2.6 cm before hitting a uranium nucleus. For a sphere well above the critical size the loss through neutron escape would be small, so we may assume that each neutron, after a life of 2.6×10^{-9} s, produces fission, giving birth to two neutrons. In the expression e^t for the increase of neutron density with time, would be about 4×10^{-9} s, very much shorter than in the case of a chain reaction depending on slow neutrons." [Let me make the same point by asking a somewhat different question but using the correct numbers. How much time does it take to fission a kilogram of ^{235}U using fast neutrons? First, the time given by Frisch and Peierls should be replaced by t $\sim$ 16.5 cm/10^9 cm/s. The number of ^{235}U nuclei in a kilogram is about 2.63×10^{24}. Thus we must solve the equation $2.63 \times 10^{24} = \exp(t/1.65 \times 10^{-8}$ s) for t which gives t $\sim$ a microsecond. Because of the fast neutrons all the explosive energy in an atomic bomb is generated in the first microsecond.]

"If the reaction proceeds until most of the uranium is used up temperatures on the order of 10^9 K and pressures of about 10^{13} atm are produced. It is difficult to predict accurately the behavior of matter under these extreme circumstances, and the mathematical difficulties of the problem are considerable. By a rough calculation we get the following expression for the energy liberated before the mass expands so much that the reaction is interrupted:

$$E = 0.2M(r^2/t^2((r/r_0)^{1/2} - 1),$$

where M is the total mass of the uranium, r is the radius of the sphere, r_0 is the critical radius and t is the time for the neutron density to multiply by a factor of e. For a sphere of radius of 4.2 cm ($r_0 = 2.1$ cm, $M = 4700$ g and t $= 4 \times 10^{-9}$ s) we find E $= 43 \times 10^{23}$ erg." [43×10^{13} J]

[There is some confusion in the literature about these formulae. I have used the expressions in the original paper which is found in the Bodelian Library in Oxford. For comparison a kiloton of TNT produces about 4.18×10^{12} J.]

"For a radius of about 8 cm, M = 32 kg, the whole fission energy is liberated according to [the equation above]. For small radii the efficiency falls off even faster than indicated by this equation because increases as r approaches r_0. The energy liberated by a 5 kg bomb would be equivalent to that of several thousand tons of dynamite and that of a 1 kg bomb, though 500 times less, would still be formidable." The efficiency of the Hiroshima bomb was 1.5%, which means that of the 52 kg of ^{235}U, only about a kilogram was fissioned. The rest floated off into thin air.]

"It is necessary that such a sphere should be made in two or more parts, which are brought together when the explosion is wanted. Once assembled, the bomb would explode within a second or less since one neutron is sufficient to start the reaction and there are several neutrons passing through the bomb in every second from the cosmic radiation. Neutrons originating from the action of uranium alpha rays on light element impurities would be negligible provided the uranium is reasonably pure. A sphere with a radius of less than 3 cm could be made up in two hemispheres, which are pulled together by springs and kept separated by a suitable structure which is removed at the desired moment. A larger sphere would have to be composed of more than two parts, if the parts, taken separately, are to be stable.

"It is important that the assembling of the parts should be done as rapidly as possible in order to minimize the chance of a reaction getting started at a moment when the critical conditions have only just been reached. If this happened, the reaction rate would be much slower and the energy liberation would be considerably reduced; it would, however, always be sufficient to destroy the bomb.

"It may be well to emphasize that a sphere only slightly below the critical size is entirely safe and harmless. By experimenting with spheres of gradually increasing size and measuring the number of neutrons emerging from them under a known neutron bombardment, one could accurately determine the critical size, without any danger of a premature explosion." [Considering the technological tidal wave that this paper was about to unleash, these remarks seem rather naive. In an actual nuclear device it is not cosmic ray neutrons that start the chain reaction but rather an "initiator" — a device that produces neutrons when the shock waves from the forced assembly of the subcritical parts strike it. These neutrons are produced when the assembly becomes super-critical. There is also something charming about the notion of this assembly being produced by the actions of "springs" naive. The kinds of experiments needed to determine the critical mass were carried out at Los Alamos, some of which by Frisch. Feynman referred to them as "tickling the tail of the sleeping dragon."]

"For the separation of ^{235}U, the method of thermal diffusion, developed by Clusius and others, seems the only one which can cope with the large amounts required. The gaseous uranium compound, for example, uranium hexafluoride, is placed between two vertical surfaces which are kept at a different temperature. The light isotope tends to get more concentrated near the hot surface, where it is carried upwards by the convection current. Exchange with the current moving downwards along the cold surface produces a fractionating effect, and after some time a state of equilibrium is reached when the gas near the upper end contains markedly more of the light isotope than near the lower end.

"For example, a system of two concentric tubes of 2 mm separation and 3 cm diameter, 150 cm long, would produce a difference of about 40% in the concentration of the rare isotope between its ends, and about a gram a day could be drawn from the upper end without unduly upsetting the equilibrium.

"In order to produce large amounts of highly concentrated ^{235}U a great number of these separating units will have to be used, being arranged in parallel as well as in series. For a daily production of 100 grams of ^{235}U of 90% purity, we estimate that about 100,000 of these tubes would be required. This seems a large number, but it would be undoubtedly be possible to design some kind of system which would have the same effective area in a more compact and less expensive form."

[Once the real work began the Clusius method was set aside. It is interesting that Frisch and Peierls are considering the design of "cascades" in which the parallel elements allow one at any stage to take on more material while the serial elements produce the separation.]

"In addition to the destructive effect of the explosion itself, the whole material of the bomb would be transformed into a highly radioactive state. The energy radiated by these active substances will amount to about 20% of the energy liberated in the explosion, and the radiations would be fatal to living beings even a long time after the explosion.

"The fission of uranium results in the formation of a great number of active bodies with periods between, roughly speaking, a second and a year. The resulting radiation is found to decay in such a way that the intensity is about inversely proportional to the time. Even 1 day after the explosion the radiation will correspond to a power expenditure on the order of 1000 kW or to the radiation of a hundred tons of radium.

"Any estimate of the effects of this radiation on human beings must be rather uncertain because it is difficult to tell what will happen to the radioactive material after the explosion. Most of it will probably be blown into the air and carried away by the wind. This cloud of radioactive material will kill everybody within a strip estimated to be several miles long. If it rained the danger would be even worse because active material would be carried down to the ground and stick to it, and persons entering the contaminated area would be subjected to dangerous radiations even after days. If 1% of the active material sticks to the debris in the vicinity of the explosion and if the debris is spread over an area of, say, a square mile, any person entering this area would be in serious danger, even several days after the explosion.

"In these estimates the lethal dose of penetrating radiation was assumed to be 1000 R; consultation of a medical specialist on x-ray treatment and perhaps further biological research may enable one to fix the danger limit more accurately. The main source of uncertainty is our lack of knowledge as to the behavior of materials in such a superexplosion, and an expert on high explosives may be able to clarify some of these problems.

"Effective protection is hardly possible. Houses would offer protection only at the margins of the danger zone. Deep cellars or tunnels may be comparatively safe from the effects of radiation, provided air can be supplied from an uncontaminated area some of the active substances would be noble gases which are not stopped by ordinary filters.

"The irradiation is not felt until hours later when it may be too late. Therefore, it would be very important to have an organization which determines the exact extent of the danger area, by means of ionization measurements, so that people can be warned from entering it."

The subject of damage from nuclear weapons is immensely complex and Frisch and Peierls barely scratch the surface. The memorandum is signed O. R. Frisch and R. Peierls, The University, Birmingham. Having written the report the question was what to do with it. They thought that it was so sensitive that Peierls typed it himself. They gave it to Oliphant who got it into the hands of Henry Tizard. Tizard was an Oxford chemist who was in charge of a committee that was studying scientific applications to wartime activities — at the time, mainly radar. They had a subcommittee that had looked into nuclear weapons, but they had decided that they were not feasible so the subcommittee was in the process of disbanding. They had considered only slow neutrons so that the Frisch-Peierls memorandum was a revelation.

Frisch and Peierls were informed that as "enemy aliens" they were to have nothing further to do with the matter. Peierls wrote a letter addressed to whoever was running whatever committee was doing this work that this position was absurd because he and Frisch knew more about this than anyone. It turned out that a new committee had been formed with the name MAUD. The reason for this name is one of the legends of the nuclear age.

Lise Meitner happened to be in Copenhagen when the Germans occupied the city in 1940. Bohr asked her to send a message to his British colleagues when she returned to Sweden. Apparently, she had no trouble getting back and wired to a friend in England: "Met Niels and Margrethe recently. Both well but unhappy about events. Inform Cockcroft and Maud Ray Kent."[16] John Cockcroft was a Cambridge physicist whom Bohr had gotten to know, but who was "Maud Ray Kent"? The recipients of the message were sure that this name was a code and that what was concealed had to do with nuclear energy. However, try as they did, they could not crack the "code." It was revealed a few years later that Maud Ray was a governess that had taken care of the Bohr children on one of their visits to England and that she lived in Kent.

In the fall of 1940 Tizard led a mission to the United States (Cockcroft and Oliphant came along) to present the results of Frisch and Peierls and the MAUD committee to various American scientists. No one was much interested. What interest there was in the use of nuclear energy for power generation and in radar, which was the central mission of the committee. However, Oliphant acted like a man possessed. He simply would not be contained when it came to discussing the prospects of a bomb. He button-holed everyone and is as responsible as anyone for getting the program revived here. There is some irony here because it was Oliphant who brought Frisch to England, which began the chain of events that finally lead to the memorandum we have been discussing.

The MAUD committee produced its final report in July of 1941.[17] It begins rather oddly. "We would like to emphasize at the beginning of this paper that we entered the project with more skepticism than belief, though we felt that it was a matter that had to be investigated. As we proceeded we became more and more convinced that release of atomic energy on a large scale is possible and that conditions can be chosen which would make it a very powerful weapon of war."[18] The body of the paper, in which Frisch and Peierls along with other prominent British scientists played a role, is one order of magnitude more sophisticated than the original Frisch-Peierls memorandum. Gone, for example, are the springs. They are replaced by high explosives — a "gun assembly" — a bomb like that of the Hiroshima weapon. Gone is the Clusius thermal diffusion method of isotope separation. It just did not work and is replaced by gas diffusion through a membrane pierced by tiny holes. Better values are available for the neutron cross sections and a modified critical mass of between 9 kg and 43 kg is presented. There are proposals to work with British industry. It is clearly a plan of action.

The MAUD committee was replaced by "tube alloys" — a code name for the British atomic bomb project. In November of 1941 Columbia University scientists Harold Urey and George Pegram attended the first meeting. They realized just how serious this nuclear weapon program was and spread the word when they got back to the United States. We can see the influence that the Frisch-Peierls memorandum had.

Both Frisch and Peierls went to Los Alamos as part of the British delegation. After the war they returned to England. In 1968 Peierls was knighted. Frisch also received several awards from his adopted country. He died in 1979 and Peierls in 1995.

References

1. Otto Frisch, *What Little I Remember*, Cambridge University Press, New York, 1979.
2. Reference 1, p. 118.
3. N. Bohr, "Resonance in uranium and thorium disintegrations and the phenomenon of nuclear fission," *Phys. Rev.* **55**, 418–419, 1939.
4. R. Peierls, "Critical conditions in neutron multiplication," *Proc. Cambridge Philos. Soc.* **35**, 610–615, 1939.
5. Reference 4, p. 610.
6. U.S. Department of Energy, "Criticality experiments facility," National Nuclear Security Administration Nevada Site Office, 2005-06-15 www.nv.doe.gov/library/factsheets/DOENV_1063.pdf.
7. Reference 3, p. 419.
8. A.O. Nier, E.T. Booth, J.R. Dunning, and A.V. Grosse, "Nuclear fission of separated uranium isotopes," *Phys. Rev.* **57** 6, 546, 1940.
9. Klaus Clusius and Gerhard Dickel, "Zur trennung der chlorisotope," *Naturwiss.* **27**, 148–149, 1939.
10. There are several places where this memorandum can be found. My choice is in an appendix of Robert Serber, *The Los Alamos Primer*, University of California Press, Berkeley, 1992. The advantage is that the reader can get a tutorial in bomb physics in the rest of the book. Also it is one of the more accessible choices. The reader should be warned that there are several significant mistakes in this version, which Cameron Reed has been able to compare to the original in the Bodleian Museum at Oxford. I will point out these mistakes as we go along.
11. I am grateful to Carey Sublette for his help in sorting this out.
12. See, for example, J. Bernstein, "Heisenberg and the critical mass," *Am. J. Phys.* **70** 9, 911–916, 2002.
13. F. Perrin, "Calcul relative aux conditions éventuelles de transmission en chaine dl'uranium," *C. R. Acad. Sci. URSS* **208**, 1394–1396, 1939.
14. L. A. Goldstein, A. Rogozinski, and R. J. Walen, "The scattering by uranium nuclei of fast neutrons and the possible neutron emission resulting from fission," *Nature London* **144**, 201–202, 1939.

15. I thank Cameron Reed who has seen the manuscript for communications on this matter. I will not present a derivation because it would take us into too much technical detail.

16. See, for example, Richard Rhodes, *The Making of the Atomic Bomb*, Simon and Schuster, New York, 1986, p. 340.

17. To read the report, see Margaret Gowing, *Britain and Atomic Energy 1939–1945*, Macmillan, London, 1964. You will also find a version of the Frisch-Peierls memorandum which is less error prone than the Serber version.

18. Reference 13, p. 594.

Part III. Mathematics and Finance

15. Bachelier

Although the name Louis Bachelier is fairly well-known to people who work either in probability theory[a] or financial engineering[b] it is unlikely to be known to physicists. Yet, in 1900, Bachelier published a thesis, *Théorie de la Spéculation*,[c] in which he anticipated results that five years later Einstein, who had surely never heard of Bachelier, re-invented and presented in his 1905 paper on Brownian motion. Einstein called his paper, "On the Movement of Small Particles Suspended in a Stationary Liquid Demanded by the

---

[a]See, for example, *An Introduction to Probability Theory and Its Applications, Volume II*, by William Feller, John Wiley & Sons, New York, 1966. On page 98 Feller introduces Brownian motion which he refers to as the Wiener-Bachelier process.

[b]For a popular account see *Capital Ideas* by P.L. Bernstein, The Free Press, New York, 1992. See, for example, pp. 18–23 and the other references in the book.

[c]A facsimile French version is available from Editions Jacques Gabay, Paris, 1995. An English translation can be found in *The Random Character of Stock Market Prices*, Ed. P.H. Cootner, M.I.T. Press, Cambridge, 1970, pp. 17–78. Cootner has supplied some useful notes, but there are misprints in some of the mathematical formulas. I would like to thank Emanuel Derman for supplying me with this translation.

Molecular-kinetic Theory of Heat,"[d] It contains the famous sentence, "It is possible that the movements to be discussed here are identical with the so-called "Brownian molecular motion"; however, the information available to me regarding the latter is so lacking in precision that I can form no judgment in the matter."[e] What I want to explore in this note is what exactly of Einstein did Bachelier anticipate. Before I do this I want to say something about Bachelier's life.[f]

Louis Jean Baptiste Alphonse Bachelier was born in Le Havre on the 11 March 1870. His father Alphonse was a wine merchant. He must have been something of an aristocrat since he was also the vice-consul of Venezuela in Le Havre. Bachelier's mother was a banker's daughter. Bachelier was surely headed for the *grande écoles* — the École Polytechnique and the like — but in 1889 both his parents died. He became head of the family business *Bachelier fils*. Soon after, he began his compulsory military service so it was not until 1892 that he was finally able to go to Paris to begin his studies at the Sorbonne. In particular, he attended lectures of Henri Poincaré who later became one of the referees for Bachelier's thesis. It seems as if Bachelier was not an outstanding student as compared to people like Paul Langevin, who took some of the same courses. Very likely the years that kept him out of school played a role. He did not have much experience taking examinations. During this period he seems to have worked at the Bourse — the French stock exchange. One would

---

[d]This paper is reproduced in the collection *Investigations on the Theory of Brownian Movement*, edited by R. Fürth and translated by A.D. Cowper, Dover Publiactions, New York, 1956.

[e]Fürth, op. cit. p. 1.

[f]The best sources I have found are Louis Bachelier, "On the Centenary of *Théorie de la Spéculation*" by Jean-Michel Courtault, Yuri Kabanov, Bernard Bru and Pierre Crépel, *Mathematical Finance*, Vol.10, No.3 (July 2000), 341–353. I am grateful to Loren Cobb for supplying this reference and for helpful correspondence and "Bachelier and his Times: A Conversation with Bernard Bru," by Murad S. Taqqu, http//www.stochastik.uni-freiburg.de/bfs.web/lbachelier/bachelier-kup1.pdf. I have enjoyed several useful correspondences with Professor Bru. *The (Mis)Behavior of Markets* by Benoit Mandelbrot and Richard Hudson, Basic Books, New York, 2004 has a chapter on the life and work of Bachelier.

imagine that it was Bachelier's experience in the world of finance that led him to do a thesis on the mathematics of financial speculation. It was sufficiently outside the norm of theses in mathematics so that it could only be given the grade of *honorable* as opposed to *trés honorable,* although the report of the referees was very favorable.[g]

It is unclear what Bachelier did to support himself after his degree. He may have continued to work on the Stock Exchange. He did manage to get a few small grants that were under the administration of Bachelier's contemporary, the very distinguished mathematician Émile Borel. Bachelier continued to publish occasionally including a long paper in 1901 which he called *Théorie Mathématique de Jeu*;[h] a treatise on the mathematics — the laws of chance — of gambling. What is striking about both this paper and the thesis are the detailed numerical calculations Bachelier made. For example, in his gambling paper he supplies a table of the Gaussian error function to six decimal places when the upper limit on the integral varies between 0 and 4.80. For the fun of it I compared a few of his answers to what I found using a modern computer program and found that they agreed to the first four places after which there was considerable scatter.

In 1909, Bachelier became an unpaid lecturer at the Sorbonne. He was being considered for a more permanent job but in 1914, he was drafted into the French army as a private and served until 1918. The first real academic job he had was as a replacement for a professor on leave at the university in Besançon. He held various temporary jobs for the next few years but his economic position must have been sufficiently secure so that in 1920, he married, although his wife died soon after. Professionally, Bachelier's great misfortune was the death of Poincaré in 1912. Poincaré was the only one of Bachelier's contemporaries who really understood what he had done. It was Poincaré who wrote the referee report on the thesis. Along with everything else, Poincaré was a master of probability theory. He could see that, despite the lack of formal rigor in the thesis, that what Bachelier had done was basically correct. He also did not have a snobbish attitude

[g]See the appendix in Courtalt *et al.* where the full report is given.
[h]Gaby, op. cit.

towards Bachelier's relatively humble educational background or the fact that his thesis dealt with the stock market and not some abstruse form of the theory of functions which was then the mode. There was no one of stature to defend Bachelier when, in 1926, he was accused of having made basic mistakes in his work. This came about because Bachelier had applied for a permanent position in Dijon. He was turned down in favor of a candidate who had all the right prestigious credentials. Bachelier was now fifty-six and his reputation was unjustifiably in tatters. The man who had done the most damage was a very distinguished mathematician named Paul Lévy. Lévy changed his mind a few years later when he discovered that, in his work on probability theory, Andrei Kolmogorov, one of the greatest mathematicians of the 20th century, made reference to Bachelier.[i] There was good reason. In some of the important results Bachelier had gotten there first by some thirty years. In 1927, Bachelier became a professor in Besançon where he remained until his retirement ten years later. He died on the 28 April 1946. This is the outline of Bachelier's life, now what did he do?

Most of Bachelier's thesis consists of applications to the stock market. The financial instruments that he dealt with have a family resemblance to the kind of "derivatives" a modern financial engineer deals with — "puts" and "calls" for example — but the French versions are different in detail and are rather confusing. Fortunately that is not what interests us here. We are interested in Bachelier's method. Here is where the overlap with Einstein is. But we can nonetheless use Bachelier's language keeping in mind that when he employs the term "price" this might be replaced in Einstein's example by "position". We can do no better than to quote this part of Bachelier's thesis, if for no other reason than to get a feel for the flavor. This is what he writes under the heading "The Law of Probability."

---

[i] Professor Bru has emphasized to me that one of the most original parts of Bachelier's this was his calculation of the curve of maximal probability per time interval by a method of reflections. Lévy discovered the same method independently only to learn that Bachelier had anticipated him by several decades. Professor Mandelbrot was kind enough to send me the text of a letter he received from Lévy explaining how he failed to see the full value of what Bachelier had done.

"The law of probability can be derived from the principle of joint probabilities.

Let $p_{x,t}dx$ designate the probability that, at some time t, the price will be included in the elementary interval x, x + dx.

We seek the conditional probability that price z will be quoted at the moment $t_1 + t_2$, the price x having been quoted at the moment t_1.

By the principle of joint probabilities, the desired probability will be equal to the product of the probability that price x will be quoted at the moment t_1, i.e., $p_{x,t_1}dx$,[j] and the probability that, given x at the moment t_1, price z will be quoted at $t_1 + t_2$, i.e., multiplied by $p_{z-x,t_2}dz$ will be[k]

$$\int_{-\infty}^{\infty} p_{x,t_1}p_{z-x,t_2}dxdz.$$

The probability of this price z at the moment $t_1 + t_2$ may also be expressed as p_{z,t_1+t_2}. Thus

$$p_{z,t_1+t_2}dz = \int p_{x,t_1}p_{z-x,t_2}dxdz,$$

or

$$p_{z,t_1+t_2} = \int_{-\infty}^{\infty} p_{x,t_1}p_{z-x,t_2}dx.$$

This is the equation which the function p must satisfy..."

Physicists who have studied stochastic processes will recognize this as a form of the so-called Chapman–Kolmogorov relation. In 1928, the British mathematician and geophysicist Sydney Chapman published a paper on Brownian motion[1] in which he derived the

---

[j] I am taking the translation from Cootner op. cit., but, as mentioned before, there are notational errors. This is one of the places and I am correcting them.

[k] When Bachelier was rediscovered by the mathematical economists like Paul Samuelson in the 1950s they noted that Bachelier had included negative stock prices. The modern analysts discuss what is called "geometric Brownian motion" for stock prices which means taking the exponential and hence the log normal distribution.

[1] On the Brownian Displacements and Thermal Diffusion of Grains suspended in a Non-Uniform Fluid, *Proc. Royal Soc. A* 119 (1928) pp. 34–60. I am grateful to Bernard Bru for this reference. The Chapman-Kolmogorov equation is equation

same equation in a non-mathematical way. It was Kolmogorov who, in 1931, published the rigorous mathematical derivation.[m] As I have mentioned, Kolmogorov made reference to Bachelier. Chapman didn't. He probably had never heard of him. Indeed, when Chapman learned that the result he had derived to solve a physics problem was now a mainstay of the theory of probability he was surprised and a little embarrassed.[n] The terminology "Chapman–Kolmogorov equation" came into use in the 1930s.

The way in which Bachelier made use of his relation appeals to a theoretical physicist, although one can understand why it would also bother a mathematician. Without giving any motivation he takes as a candidate a Gaussian in the variable x. He very likely had the central limit theorem in mind, something he had learned in Poincaré's probability course. He does not discuss the uniqueness of this choice. In fact, it is far from unique. For example the gamma distribution given by

$$(1.) \qquad f_{\alpha,\nu}(x) = \frac{1}{\Gamma(\nu)}\alpha^{\nu-1}e^{-\alpha x}$$

is also closed under convolutions.[o] Bachelier makes no attempt to address this issue. What he does is to assume that the function $p_{x,t}$ takes the form

$$(2.) \qquad p_{x,t} = p(t)\exp(-\pi p(t)^2 x^2).$$

---

four of Chapman's paper. He gets new terms in his expansions because the fluid is not uniform. When he restricts himself to the uniform case the Gaussian solution looks unfamiliar because he is dealing with three dimensions.

[m]Über die analytischen Methoden in der Wahrscheinlichkeitsrechnung, *Math. Ann*, 104 (1931) pp. 415–458. Again I am grateful to Professor Bru for this reference.

[n]See D.G. Kendall, *Bulletin of the London Mathematical Society*, Vol 22, Part 1, January 1990, p. 35.

[o]Feller op. cit. p. 46.

He then shows by simple integration that with times t_1 and t_2, whose corresponding p's I will designate as p_1 and p_2, we have[P]

$$\int_{-\infty}^{\infty} p_1 \exp(-\pi p_1^2 x^2) p_2^2 \exp(\pi p_2^2 (z - x)^2) \, dx$$

$$(3.) \qquad = \frac{p_1 p_2}{\sqrt{p_1^2 + p_2^2}} \exp\left(\frac{-\pi p_1^2 p_2^2}{p_1^2 + p_2^2} z^2\right)$$

Equation (3.) shows that Gaussians go into Gaussians under convolution but it does not deal with the time behavior. Bachelier handles this with a very clever bit of functional analysis. Let us define $f(p_1, p_2; t_1, t_2)$ by the following

$$(4.) \qquad f(p_1, p_2; t_1, t_2) \equiv \frac{p_1 p_2}{\sqrt{p_1^2 + p_2^2}}.$$

We are going to find a condition on the p's so that f is a function of $t_1 + t_2$. The condition that we are going to use is that if this is true we must have $\frac{\partial}{\partial t_1} f = \frac{\partial}{\partial t_2} f$. We will now apply this to f^2. If we do this we are led to the equation

$$(5.) \qquad \frac{\frac{d}{dt_1} p_1}{p_1^3} = \frac{\frac{d}{dt_2} p_2}{p_2^3},$$

or

$$(6.) \qquad \frac{\frac{d}{dt} p(t)}{p(t)^3} = C.$$

Here C is an constant. This equation integrates to

$$(7.) \qquad p(t) = \frac{C}{\sqrt{t}} + C'.$$

Without discussion Bachelier sets C' equal to zero. He probably knew the answer he wanted to get when $t = 0$, or maybe he employed the

---

[P]There is an important misprint in Cootner op. cit. which I have corrected.

unstated initial condition that at t = 0 the probability is concentrated at x = 0. In any event he arrived at

$$(8.) \qquad p_{x,t} = \frac{C}{\sqrt{t}} \exp\left(-\frac{C^2}{t}\pi x^2\right).$$

A physicist looking at this equation will have an immediate sense of *déjà vu*. In Einstein's theory of Brownian motion it describes the diffusion of the "small particles." Indeed, it is Equation (10.) of his paper. I will explain how Einstein derived it shortly but I want to continue on a bit with Bachelier.

Much of the rest of the thesis is then devoted to financial applications. He defines something he calls the "positive expectation" of price as $\int_0^\infty p_{x,t} x dx$ and shows that it goes like $\sqrt{t}$. Later in the thesis he gives a second derivation of the Gaussian distribution. I will only sketch it briefly because it does not have any overlap with Einstein's paper. He supposes that there are two mutually exclusive events A and B, with respective probabilities p and q = 1 − p. He then wrote down the probability that after m events α is equal to A and $m - \alpha$ equal to B, i.e., $\frac{m!}{\alpha!(m-\alpha)!} p^\alpha q^{m-\alpha}$. He then computes the expectation of a gambler who would wager the amount of the "spread" away from the maximum probability. He finds an expression for this which he approximates using Stirling's formula for large values of the factorial. The spread h is then in an interval between h and h + dh with a probability $\frac{dh}{\sqrt{2\pi mpq}} \exp(-\frac{h^2}{2mpq})$. To complete the argument he supposes that the time interval t is divided into m intervals of width Δt, i.e., $t = m\Delta t$. To make contact with his previous result he makes the random walk assumption that p = q = 1/2. In the context of financial speculation this is the assumption that in a "fair market" at any given time irrespective of past history the price of a stock is equally likely to go up as down. This assumption underlies much of modern financial engineering and has its share of critics.[q] Incidentally, one of the criticisms that Poincaré made of the thesis in his report was that the assumptions that went into Bachelier's derivations were not made more explicit.

---

[q]For an engaging example see Mandelbrot and Hudson op. cit.

There is a part of the thesis that again gets us close to Einstein. Bachelier discusses something he calls the "radiation" or "diffusion" of probabilities. He very clearly has in mind the diffusion of heat. I find the details of what he does here somewhat confusing. It is probably his language. The essence is that he supposes that in what he calls an "elementary time interval" a price x radiates or diffuses an amount of probability that is proportional to the difference of probabilities for x and for its neighboring price. He certainly has in mind Fourier's treatment of heat diffusion as being proportional to temperature differences. He defines **P** as the probability that a price can be found at a value x at a time t. Although he does not use this language, the probability current is then $-\frac{d\text{P}}{dx}$. Then, basically invoking what we would call the equation of continuity for this current, he arrives at the equation

$$(9.) \qquad c^2\frac{\partial \mathbf{P}}{dt} = \frac{\partial^2 \mathbf{P}}{\partial x^2}.$$

Here c is a constant put in for dimensional reasons. He calls this the Fourier equation. Although he does not explicitly say so, he must have known that the Gaussian normal distribution, Eq. (8.), is a solution to this equation, something that would have completed his discussion. We can now turn to Einstein.

The part of the 1905 paper I want to focus on is section four entitled "On the Irregular Movement of Particles Suspended in a Liquid and the Relation of this to Diffusion." In short, Einstein is dealing with a physics problem and not an exercise in abstract probability theory. His basic idea is to introduce a time τ which is short compared to the total time of the observation but of such a size that movements of the suspended particles in any two such consecutive intervals can be considered as independent events. He supposes that there are a total of n particles in the suspension. In the interval of time τ the x-coordinate of any one of the particles will increase or decrease by an amount Δ. The Δ's are distributed by some probability law $\phi(\Delta)$ following Einstein's notation. This function has the properties

$$(10a.) \qquad \phi(\Delta) = \phi(-\Delta),$$

and

(10b.) $$\int_{-\infty}^{\infty} \varphi(\Delta)d\Delta = 1.$$

In the time interval τ the number of particles dn that receive a displacement between Δ and $\Delta + d\Delta$ is given by

(11.) $$dn = n\phi(\Delta)d\Delta.$$

He defines $f(x, t)$ as the number of particles per unit volume in the neighborhood of x at the time t. He then wants to see how the distribution appears at a time $t + \tau$. He confines himself to a region of width dx between two planes that meet the x-axis at right angles at points x and x + dx. Thus[r]

(12.) $$f(x, t + \tau)dx = dx \int_{-\infty}^{\infty} f(x + \Delta, t)\phi(\Delta)d\Delta.$$

This is a special case of the Chapman–Kolmogorov–Bachelier equation. To deal with it Einstein expands the left-hand side in τ and the f under the integral in Δ. The zero-order terms cancel using the normalization of ϕ. The odd powers of Δ under the integral vanish because ϕ is an even function. We are left with

(13.) $$\frac{\partial f}{\partial t} = D\frac{\partial^2 f}{\partial x^2},$$

where

(14.) $$D = \frac{1}{\tau}\int\frac{\Delta^2}{2}\phi(\Delta)d\Delta.$$

Einstein now changes coordinates to a system which at t = 0 has an origin at the center of gravity of the particles. This is the problem of the particles diffusing outwards from a point with the initial

---

[r]In the English translation of this paper that I am using, Fürth op. cit. the time dependence of the function f under the integral sign is left out. I do not have the German original to see if it is left out there as well. In his paper Chapman is careful about all these dependences.

condition that the particles are concentrated at x = 0.[8] This leads to the solution

(15.)
$$f(x, t) = \frac{n}{\sqrt{4\pi Dt}} \exp\left(-\frac{x^2}{4Dt}\right).$$

The most famous equation in the paper follows from this, i.e.,

(16.)
$$\sqrt{\langle x^2 \rangle} = \sqrt{2Dt}$$

from which Avogadro's number was independently determined.

As far as I know Bachelier never mentioned Einstein in his work. Einstein certainly had never heard of Bachelier, who was rediscovered only in the 1950s by the economists. I sometimes ask myself if they, had met, what would the two men have had to say to each other. I suspect very little.

---

[8]In modern language the limit when $t \to 0^+$, $f(x, t)$ is the function $\delta(x)$.

16. A Primer of Things You Need to Know to Follow the Financial News

"Financial crisis is by definition something that had not been anticipated. If it had been anticipated it could have been arbitraged away."

Alan Greenspan

At the outset you need to know that I am a theoretical physicist and
not a stock broker or commodities trader. But I have been trying
to follow the news. It is full of terms such as "mark to market" and
"credit default swaps" which at first meant about as much to me as
"Bell's Inequality" might mean to some of you. But I am stubborn
and decided that if these things are going to determine my financial
future I had better know what they mean.

The first thing that becomes clear when you look into the matter
is that these concepts are part of a virtual economy. They are not
like the "real economy" where, after you go into a grocery store to
buy eggs and milk, you come out of the store with eggs and milk
and not an option to buy eggs and milk at an agreed price at a
later time. If you buy a "credit default swap" you come out of the
transaction with a piece of paper — or more likely a computer file —
which has no intrinsic value. If you imagine a situation in which you
are confronting an aboriginal tribe you might get somewhere if you
have beads to swap, but not if you have a credit default swap. These
financial instruments are rather new. The commodity they trade in
is money. They are very clever devices that were thought up by very

smart people to make money from money and they are in the process of doing us all in. I am somehow reminded of something that the great German mathematician David Hilbert said about astrology. If the ten smartest people in the world got together they could not think of anything as stupid as astrology. Now to the glossary.

1. Derivatives 101

This is such a complex subject, and so fundamental, that I am dividing this entry in the primer in three parts. This one, 101, gives the basics.

A derivative is a financial instrument whose value is derived from the values of other financial instruments that do have value.[a] For example if I buy an option to buy some stock in the future at some present price, what determines the value of the option depends on the future value of the stock. The question is, what is such an option worth to me now? To understand this better let us consider a specific example.

Suppose you are given the following information — in reality you are never given this much information but that is for Derivatives 201. You are told that the Horse Feather company, whose stock is presently worth $100 a share, in six months will be worth $120 with a probability $\frac{3}{4}$ or $80 with a probability $\frac{1}{4}$. You are offered an option to buy this stock in six months at $100 a share. What should you pay for this option? This form of option is called a "European" option since it can be exercised only after six months. A so-called "American" option can be exercised any time. Your expectation for a gain after six months is, in dollars, $3/4 \times 20 + 1/4 \times 0 = 15$ which reflects that fact if the stock drops to $80, the option is worth nothing. This calculation appears to show that the option is worth $15. But never underestimate the ingenuity of people interested in making money. In this case there is the concept of "arbitrage."

Let us suppose you are willing to buy the option from me for $15. I will take the $15 and pocket $5 which you will never see again. I then

---

[a]Emanuel Derman made the point to me that money itself is a kind of derivative. It used to be linked to gold and now to other material things.

go to my friendly bank and borrow $40. This is the "leverage". In all these transactions I am assuming that there are no stock commissions and no bank interest. Since the people who thought these things up often had a background in physics they called these transactions "frictionless." It is not difficult to take the "frictions" into account. I take the $40 and the $10 and buy a half share of the Horse Feather company. (If you object to half shares I can modify my example to make it full shares.) This purchase of the half share is known as the "hedge." This sort of thing raised to many powers is what "hedge funds" do. I will now show you that I can't lose.

There are two cases the final price is $120 or the final price is $80. In the former case you will exercise your option which was our agreement. But you are not interested in owning the stock, but simply pocketing the twenty dollar profit which I owe you. I will now sell my half share for $60. Forty of this I will give back to the bank and the remaining twenty I give to you. Note two things. In the first place I get to keep the five dollars. In the second place the true cost of the option was ten dollars since the rest was money borrowed from the bank. You overpaid for the option by five dollars. Now take the second case. The stock has dropped to $80. You do not exercise your option so I owe you nothing. But I sell my half share for $40 which I give back to the bank still pocketing the five dollars.

This seems too good to be true and it is. In Derivatives 201 I will begin to discuss the real world.

2. Derivatives 201

In 101 I presented a "toy model" to illustrate the basics. I ignored the "friction" of such inconveniences as broker commissions and bank interest. But, by somewhat complicating the mathematics, these can readily be taken into account. What makes the model a "toy" is the assumption that we know the possible future values of the stock and the probabilities of the stock having one of these values by some sort of clairvoyance. In the real world one replaces clairvoyance by mathematical modeling, although the way things have turned out one often longs for a spirit medium.

The idea of such a mathematical model goes back to the year 1900 when Bachelier published his PhD thesis *Théorie de la Spéculation*.

Bachelier was already thirty years old and had spent time working at the *Bourse* — the French stock exchange. The question he asked was precisely the one that interests us; How can we predict the future probable values of a stock given our present information? To answer this he proposed the idea that stocks follow a "random walk." As far as I know he never later connected this to a problem in physics solved in 1905 by Einstein. Einstein surely had never heard of Bachelier. Very few people had until he was rediscovered in the 1950s by mathematical economists like Paul Samuelson.

The problem in question was stimulated by a discovery that the Scottish botanist Robert Brown made in 1827. He noticed that if microscopic pollen grains were suspended in water then these grains performed an odd random movement which is now fittingly called "Brownian motion." At first Brown reasonably thought that these grains might be alive which explained the motion. But he tried all sorts of other suspended particles including soot from London and all of them exhibited the same motions. Throughout the 19th century this remained a puzzle although the correct solution was conjectured; namely that the suspended particles were being bombarded from all sides by the agitated and invisible molecules belonging to the liquid in which the particles were suspended. It is amusing that one objection to this is one that people first exposed to this idea often make. Such people argue that if the particle is being bombarded from all sides how does it get anywhere? But after the first lurch in some direction it is infinitesimally probable that the second one will reverse the first. The particle will go off in a new direction. It was Einstein who in 1905 made all of this quantitative. He was able to show that the average distance such a particle would travel from its origin in a time t was proportional to the square root of t which could and was tested experimentally.[b] As I mentioned, I find no reference to Brownian motion in anything that Bachelier wrote.

---

[b] For the fastidious reader I note that what I have called the "distance" here is the square root of the mean square distance and that Einstein's reasoning was somewhat different. The Polish physicist Marian Smoluchowski at about the same time analyzed Brownian motion as a random walk.

The problem that Bachelier posed was suppose you know the price of some stock now, what is the probability that the stock will have some given price at a later time? He did this by supposing that the later probability was achieved with a number of steps in which any motion of the stock was equally likely to be up or down. This random walk implies an "efficient market". In this kind of market prices are set by market conditions and that fluctuations do not matter. No scheme will help you to beat the market. In fact all market speculation is based on the assumption that at least in the short run this is false. As a student of probability theory, Bachelier understood that with his assumptions the probability, as the number of ups and downs increased, would approach a "normal" distribution — a bell-shaped curve. From this he could predict the most likely future price. Here we must mention something crucial for what has happened. A property of the bell-shaped curve is that it has "wings". No matter how far you are away from the most likely there is always a non-zero probability for something that is very unlikely. You may say that it is irrational to worry about the unlikely but keep in mind what Lord Keynes noted, "The market can remain irrational longer than you can remain solvent."

Bachelier applied his methodology to derivatives — options, but one thing did not seem to have occurred to him — arbitrage — beating the system by hedging. This was corrected in the 1970s primarily by three mathematical economists, Myron Scholes, Fischer Black and Robert Merton. Merton and Scholes were then at MIT. While Black was a consultant for the Arthur D. Little Company. Scholes and Merton shared the 1997 Nobel Memorial Prize in Economics. Black had died two years earlier. The work was done independently by Black and Scholes, and by Merton. The mantra of the quantitative financial analysts — the "quants" as they are called — is the so-called Black–Scholes equation. It is derived in a way that Bachelier would have understood. The same sort of assumptions about Brownian motion are used. The solutions tell you how to price options. For a scientist like myself it is a curious affair. In quantum mechanics, to take an example, using the mathematics of the theory we can predict the probable results of future experiments from present data. Here we use the probable future to retrodict the present. In quantum mechanics we cannot even describe the past. There are several possible pasts with different probabilities.

Merton's approach was different and it is the one that is most commonly used by people who deal in these things. He showed that the actual option could be replaced by a "synthetic" option consisting of stocks and cash that reproduced the results of the real option. In fact you need never make reference to the real option. We saw this in the toy model. The stock purchased using the borrowed cash along with the ten dollars replicated the option. Once this was understood the quants had a field day. This was especially true since, as market conditions changed, the mixture of stocks and cash had to be continually adjusted. This could not be done by hand. It had to be done by computers. Dealing with derivatives was like dealing with a black hole. Moreover there was no *a priori* check on the model. You only knew that it didn't work when it led to a financial disaster. This happened in 1987 and again in 1998 and again in 2008. It is the subject of the next entry — 301.

3. Derivatives 301

John Meriwether was born in Chicago in 1947. In his teens he won a caddie scholarship — a scholarship only open to caddies — which he used to attend Northwestern University. After a year of postgraduate teaching he went to the University of Chicago to study business. One of his classmates was Jon Corzine. In 1973, Meriwether took a job at Solomon in New York. This was just before the financial engineering revolution. After it took over, Meriwether formed the arbitrage group at Solomon. His modus operandi was always the same. He looked for the smartest people he could find even if they were smarter than he was. It did not matter how gooky they were. My feeling about Meriwether was while he certainly liked to make money he was much more interested in showing that he and his people we smarter than anyone.

In 1994 Meriwether founded a hedge fund named Long Term Capital Management — LTCM — in Greenwich, Connecticut. He hired, among others, Merton and Scholes. They don't get any smarter than that. They brought to bear every bit of wizardry you could get out of computer models. For awhile it worked marvels. It was scoring returns of 40% for its investors. It had a trillion dollars under management. This was more than any of the large investment banks. But by four years later the bubble had burst and in less than four months in 1998 they lost close to five billion dollars. What was worse was that they were not playing with their own money. They were leveraged to the hilt which might have been all right if the institutions that had lent them money had not wanted it back once they saw that the ship had hit an iceberg. In the event, LTCM did not have the money to repay these loans. In fact Meriwether tried to borrow more on the theory that, given a bit more time, the hole in the side of the ship would repair itself. What happened?

In the three disasters I mentioned there is a common element. Some financial instrument was created that could not fail so long as the market functioned rationally. The wings of the bell curves didn't matter. As a physicist I think of these things in terms of thermal equilibrium. If you have a gas at some temperature you can predict what the average energy of the molecules that compose it has. But there are fluctuations — deviations from the average. Under most circumstances these fluctuations dissipate. If there was a circumstance in which they didn't we would be in hot water at least for awhile. In the case of LTCM the instrument was what is known as "convergent trades." I will discuss the ones that characterize the other debacles in due course. First convergent trades.

Let us take a simple example. Suppose you have two treasury bonds, one that matures in thirty years and one that matures in twenty-nine years. Let us further suppose that we spot the fact that the second bond is trading for somewhat less than the first. It is reasonable to assume that this is a temporary fluctuation and that the spread between the two bonds will approach zero as the bonds mature and the price of the cheaper bond rises and that of the more expensive bond declines. Here is how we play the game. We "short" the expensive bond. We borrow the shorted bonds from some willing bank, or what have you, and then immediately sell them. If, as expected, the price of this bond drops as the two converge we can then replace the bonds we borrowed at less cost. But suppose instead the price of this bond rises and the cost of covering the short then

increases. If this is a brief fluctuation then we can sit tight until things
sort themselves out and behave as they are supposed to. But suppose
they don't and the spread gets wider and wider? Then, not to put
too fine a point on it, our goose is cooked. This is what happened to
LTCM.

The first intimation of trouble was in Thailand in the summer of
1997 with the collapse of the currency which caused people in the
Pacific Rim countries to look for investments such as our treasury
bonds that seemed more secure. This turned into a panic when Russia
stopped payment on its debt payments in August of 1998. There was
a flight for safety everywhere. LTCM had been using its convergent
trade strategy everywhere. They had even opened an office in Japan.
The spreads kept getting wider and LTCM more desperate. The obvious response would normally have been tough nuggets. But LTCM
was in hock to the tune of about a hundred and twenty **billion** to
its lenders — some of the most prestigious and important financial
institutions in the country. In light of what has happened it is interesting to recall the role of Bear Stearns in this. Bear Stearns was the
broker of record for LTCM. They kept a reserve of LTCM assets —
"cash in the box". The condition was that if this reserve fell below
500 million Bear Stearns would no longer trade for LTCM and the
party would be over. By September when it became clear to Bear
Sterns that the assets of LTCM were declining they asked for an
additional 500 million. Meriwether tried to borrow more money from
everyone including his old classmate Jon Corzine at Goldman Sachs
all to no avail. One problem was that as an unregulated enterprise
LTCM's books were closed and no one from the outside could find
out the details.

The dénouement began on Sunday, September 20 when represen-
tative of the New York Fed along with bankers from Goldman Sachs
and JPMorgan made the short trip from New York to Greenwich to
examine the condition of LTCM first-hand. It turned out that these
bankers had no inkling of LTCM's off-book trading strategies even
though they were counterparties to billions of dollars in loans. It was
also clear that LTCM was too big to fail. That would bring down a
number of other major financial institutions and the financial structure might collapse like a house of cards. The portfolio of LTCM had

to be bought out in a fire sale. One of the bidders was none other than Warren Buffet who attached very stiff conditions such as the firing of the entire management team of LTCM. Meriwether rejected this offer and in the end, 13 banks bought LTCM out and closed the fund.

Before I turn to the other two cases — 1987 and the present — it will be helpful if I add a few more primer entries. They will come up later.

4. LIBOR

Some years ago I made a small investment in a fund offered by Merrill Lynch. I was informed that the returns were adjusted to something called LIBOR. When I asked what that was I was told that it was short for London Interbank Offered Rate. How you get "LIBOR" out of this I am not sure. Though I had no idea why a London offered bank rate should have anything to do with much of anything, I did not at the time have the intellectual curiosity to inquire further. I contented myself by going around and saying that I was in Libor — which would have been marginally more funny if I had been Australian. I have long ago sold the fund and would have given it no more thought except that LIBOR has come back big time in the present economic crisis. This has motivated me to look into the matter.

I was surprised to learn that LIBOR is a fairly recent institution. It had begun informally in 1984 but only became official on January 1, 1986. It was a response to the fact that a variety of new financial instruments had appeared with a variety of different interest rate policies. It was thought that it would be a good idea to bring some uniformity to the process. In essence there are sixteen London banks which supply by 11 a.m. London time to a central office the rate at which they could borrow money, based only on their own assets as collateral, from other banks at that time. How this is determined is as much an art as a science because the banks in question do not have to have made such a transaction.

To a suspicious mind like my own, the first question that occurs is that if these rates are used to set a variety of other rates all over the world, would not some of these sixteen banks be tempted to put a little "body English" on their numbers. The tiny staff that receives these numbers at an office in London's Docklands looks for anomalies. Moreover when it averages them to produce the daily LIBOR it throws out the highest and lowest number it has gotten from the banks. Nonetheless. Why has the LIBOR come to special prominence now? This has to do with its relation to the federal funds rate. This is a rate that is set monthly by the Federal Reserve Open Market Committee. It determines the rates at which American banks in the system will lend money to each other and the rate at which the fed will lend to member banks. Its purpose is normative. Raising the rate will cool the economy while lowering it will in principle do the opposite. The LIBOR, like the canary in the mine shaft, simply reports. When things are normal the federal funds rate and the LIBOR track each other with the funds rate being a percent or so lower than the LIBOR. But recently the two have gotten out of alignment. The funds rate, is as of this writing 1.5%, while the LIBOR is over four. This reflects the reluctance of these London banks to lend to each other. It is a signal that something is seriously wrong. It is now clear that something was wrong. There was collusion and the banks that were caught up have been fined a fortune.

5. Credit Default Swap

A "credit default swap" is a form of insurance pure and simple-well impure and not so simple. It is called a "swap" rather than an insurance policy because that way it is exempt from regulation unlike an insurance sale. To understand the magnitude of these transactions note that worldwide the value of the loss that is covered is estimated to be some fifty five **trillion** dollars! This is about equal to the gross domestic product of the **world**! But they do not add a scintilla of productivity to anyone. They are simply financial instruments for making or losing huge sums of money fast. The buyer of a credit default swap pays the seller an amount of money to insure against the default of something like a mortgage — or a sliced and diced bundle of mortgages. But these swaps are themselves tradable. Since the whole market is unregulated no one knows who owns what until there is a default and someone has to pay up. Since the market is unregulated there is no specified amount that the sellers of these swaps have to keep on hand to pay the piper in case of default. A run on the swaps such as what is occurring in the mortgage market can put the counterparties into bankruptcy. This is what was about to happen to the American International Group (A.I.G.) which had about $440 billion in outstanding swaps for which they were responsible. The government bailed them out and several executives of the company celebrated by going on a very expensive partridge hunt in England. Let them eat partridge. AIG remains in business.

6. Mark to Market

"Mark to market" sometimes called "mark to the market" is an accounting protocol that seeks to apply the same accounting methods to other financial instruments that are routinely applied to stocks. If you own a stock you know that after the market closes its value is posted. This is marking the value of the stock to the market. If someone wanted to know your net worth and you wanted to include this stock this is the value you would give and not some possible value six months from now. But what if the value of the stock were undefined or ambiguous? Then if you were a clever and not particularly ethical accountant you could mark to the market of expected future earnings and inflate the present value of your enterprise. This is something that was done by Enron in spades. Conversely suppose that you held instruments that you were sure would increase in value in the future, but were distressed now, then mark to the market accounting might show that you were going bankrupt even though at some future time you might recover. This sort of thing was one of the elements that put Lehman Brothers out of business.

7. The Crash of 1987

The 1987 stock market crash, with its Black Monday on October 19th in which the Dow lost 22.6% of its value, is a perfect model of what can go wrong when very smart people do not think things out clearly. In this case the proximate cause was what is known as "portfolio insurance." To illustrate this let us return to the toy model. Suppose I own a share of Horse Feather which is now worth $100 and is paying a very nice dividend. I want a scheme by which I can hold the stock for awhile with absolutely no risk of loss. Here is what I can do. I can short one share. I borrow a share from my friendly broker and sell it for $100. Now after the time of interest there are two possibilities; either the share will be worth $120 or it will be worth $80. Let us consider the first case first. I have to return one share to the friendly broker. I can if I want to be a little long-winded, sell my share for $120. Take the $100 I have and add $20 and buy a new share to give back or I can just give the broker the share I have. Now let us analyze the second case. I take my $100 and use $80 of it to buy a new share which I give to the broker. I now have $20 plus one share worth $80. In either case I have lost nothing. The insurance worked marvels. What could possibly go wrong?

The underlying assumption in the scheme is that the short selling people have to do to cover their positions, will not influence the market. But so many people had jumped on this insurance that once the market began to fall, and they had to sell into this depressed market, they depressed the market further — a textbook example of negative feedback. It seems never to have occurred to geniuses that thought up this scheme that such a thing would be possible.

8. The Crisis of 2008

We are still too close to this to see how it is going to end. There is no doubt that it began with the collapse of the housing bubble. There is a point I would like to make that I have not seen much discussed. Suppose instead of housing we were talking about, say, a collapse in the price of tulip bulbs. Then if the normal laws of supply and demand are followed, then as the number of tulip bulbs decreases, assuming that people still want them, the price will start to go up. But housing is different. As the prices go down people get into more and more trouble with their mortgages. Hence there are more foreclosures and the stock of available housing **increases** depressing the prices still further — again a case of negative feedback. Only now is the housing market recovering. The stock market is booming.

Part IV. The Higgs Boson

17. A Question of Mass

> *"Definition I.* The quantity of any matter is the measure of it by its density and volume conjointly... This quantity is what I shall understand by the term *mass* or *body* in the discussions to follow. It is ascertainable from the weight of the body in question. For I have found by pendulum experiments of high precision, that the mass of a body is proportional to its weight; as will hereafter be shown"
>
> Isaac Newton[a]

When I took freshman physics as a sophomore at Harvard in 1948 this definition of mass was still used in our textbook. As it happened the previous year I had taken a kind of philosophically oriented course in modern physics given by the philosopher-physicist Philipp Frank. He had introduced us to the work of his fellow Austrian philosopher-physicist Ernst Mach. Therefore I knew of Mach's devastating critique in his book *The Science of Mechanics* and how it had influenced Einstein. The definition is, of course, perfectly circular. What is density? Moreover how does it apply to the photon which has no mass?

When I began writing my PhD thesis in the early 1950's I would have described myself as an "elementary particle" theorist as opposed

---

[a]This translation from the Latin can be found in *The Science of Mechanics*, by Ernst Mach, Open Court, Lasalle, Illinois, 1960, p. 298.

say to a nuclear theorist. Elementary particles were not supposed then to be made out of anything else, whereas the atomic nucleus was made out of neutrons and protons which were taken to be elementary particles. There were not that many elementary particles known at the time. In addition to the neutron and the proton there was the photon, the electron and the neutrinos. The positron was known and most physicists — Feynman being a notable exception, believed that the anti-proton would be found once the accelerators were energetic enough. A few years earlier a "heavy electron" had been discovered which had the properties of the electron, including its weak and electromagnetic interactions, except that it was about two hundred times more massive and unstable. For various reasons that no longer make any sense it was called the "mu-'μ'-meson" or simply the "muon." It seemed to serve no purpose and when I.I. Rabi heard of it he asked, "Who ordered that?"

Rabi's pique was understandable. In the 1930's a theory of the nuclear force had been proposed. It had to account for two things at least. First the force was very short-ranged. It acted only when the neutrons and protons were practically on top of each other. Secondly it had to be much stronger than the electrical force otherwise the positively charged protons that repel each other electrically would tear the nucleus apart. As it is, heavy nuclei with many protons tend to fission spontaneously. Both of these conditions could be met if a fairly massive particle was exchanged between the neutrons and protons and among themselves. The strength of this interaction could be postulated to be large as compared to the electrostatic force. It could also be shown that the range of this force was related to the mass of the particle being exchanged. From the uncertainty principle for energy and time taking the energy uncertainty to be this mass m and calling r the range then

$$mc^2 \sim \hbar c/r \sim 197\,\mathrm{Mev\ fermi}/r$$

where a fermi is 10^{-15} meters and is about the range of the nuclear force. Its mass was thus predicted to be about four hundred times larger than that of the electron. This mass unlike the mass of the muon seemed to have a connection to the dynamics. The muon had something like the correct mass but it only interacted electrically so it was the wrong particle. The right particle was called the "pi-meson" or the "pion." Why did it not show up in cosmic rays rather than this dreadful muon? The answer turned out to be very simple. The pi meson, when it was not absorbed in the atmosphere, decayed into a muon and neutrinos. When the accelerators got sufficiently powerful they produced pions in droves. At the time I was writing my thesis, pi meson physics was flourishing. But then the roof fell in.

Particles no one anticipated had begun to show up also in droves in cosmic rays. They became known as "strange particles" because they were. There was what was the K meson that came in a charged and neutral variety and there were what became known as "hyper-ons" which had masses greater than the proton or neutron. In this category there was a lambda which is neutral, a sigma with charges plus, minus and zero and a xi with charges zero and minus. But these were the lowest mass particles which were iterated in higher mass replicas. It was in short a zoo. It took strong nerves to see any pattern here, a pattern that might reflect an underlying symmetry.

At the time there was a clear idea of how such a symmetry might appear. The neutron and proton were prime examples. They had many things in common including their spins which were identical and their masses with did not differ by much. The neutron was a bit heavier and decayed into a proton, and electron and an anti-neutrino. Suppose one imagined a world in which electromagnetism was switched off. In this world, one imagined, the neutron and proton would have the same mass and would collapse into a kind of doublet. The three pi mesons — plus, minus and zero charge — would collapse in to a kind of triplet. A symmetry would emerge which was called "isotopic spin" — technically invariance under the group SU(2). The predictions that this led to were rather sound so isotopic spin was a useful approximate symmetry. Maybe something analogous could be found for the strange particles.

The trouble was that the mass differences were too great. While the mass difference between the neutron and proton was only a fraction of a percent of the mass of either particle, the K mesons had nearly four times the mass of the pions to take an example. It took an act of faith to see how these objects fitted into some kind of symmetric structure. But Murray Gell-Mann took the leap. He proposed a symmetry that was a generalization of isotopic spin — SU(3) — and suggested that if one ignored the various mass differences then the known particles could be organized in multiplet structures. For example the known scalar mesons including a newly discovered one that was called the "eta" fitted into an octet. The known hyperons fitted into a ten-fold decuplet but there was one missing. It was given the name Ω^- and its properties were predicted. When it showed up with these properties the lingering doubts about this scheme vanished and Gell-Mann was awarded a well-deserved Nobel Prize. It was a textbook example of a symmetry and its breaking.

This kind of symmetry breaking is nearly as old as quantum theory itself. Eugene Wigner and Herman Weyl, for example, studied the role of group theory in quantum mechanics. The idea was that the description of a quantum mechanical system was split into two parts. There was a Hamiltonian that exhibited the symmetries of the group to which was added a second Hamiltonian that did not. If this was "small" then some aspects of the original symmetry would still be apparent.

Take as an example the group of rotations in three dimensional space — SO(3). These rotations are generated by the orbital angular momentum. Suppose one part of the Hamiltonian contains only a central force. This is invariant under rotations which means that the angular momentum operators commute with this part of the Hamiltonian. The eigenstates are both eigenstates of the energy and of the angular momentum. If, for example, the nuclear force that binds the neutron and proton together is represented by a central force then the ground state of the deuteron would be an S-state. But it isn't. It has a small percentage of D-state which manifests itself in the fact that the deuteron has a quadrupole moment. The rotation symmetry is broken in this case by adding a "tensor force." The total angular momentum which includes the spin is conserved but the purely orbital part is not. Nonethless it is still useful to expand the wave functions in eigenfunctions of the angular momentum. In the isotopic spin example, the neutron–proton system in the absence of electromagnetism shows symmetries under the group of special unitary transformations, SU(2). Once electromagnetism is turned on the symmetry is broken but nonetheless there are still some useful manifestations.

Likewise the elementary particles in the absence of symmetry breaking show invariances under the special unitary group SU(3). When this symmetry is broken it is still possible to derive relationships among the masses but their origin is still unexplained. However in the early 1960's Yoichiro Nambu and others showed that a second kind of symmetry breaking was possible in quantum mechanics. This was called "spontaneous" symmetry breaking. To see what it means let us consider an example that has nothing to do with quantum mechanics.

Consider the equation

(1.) $$d^2 f(x)/dx^2 = x^2 + cx.$$

If c = 0 this equation is symmetric under x inversion; x → −x. Dropping the integration constants we have

(2.) $$f(x) = 1/12x^4 + c/6x^3.$$

The solution with c not equal to zero is not x inversion symmetric. This is in the spirit of Wigner–Weyl symmetry breaking. But consider

(3.) $$d^2 f(x)/dx^2 = x^2.$$

The equation is now x-inversion symmetric. But the solution is

(4.) $$f(x) = 1/12x^4 + Bx + C.$$

Unless B is zero the solution does not have the same symmetry as the equation. The symmetry has been "spontaneously broken" by the choice of solution, the initial conditions. To see how this works in quantum mechanics and to appreciate the consequences we consider a pertinent example: the self interactions of a complex scalar field,

$$\Phi(x,t) = \Phi(x,t)_1 + i\Phi(x,t)_2$$

where $\Phi_{1,2}$ are real fields. This describes a charged spinless particle. It is the simplest example I know of and it is the one that was first considered historically. It will lead us to the Higgs mechanism. We begin by exhibiting the Lagrangian of a **free** complex scalar classical field which corresponds to a particle that has a mass m. To make the notation more compact I will employ the usual convention of setting c = 1. Thus

$$L = \partial_\mu \Phi(x)^\dagger \partial_\mu \Phi(x) - m^2 \Phi(x)^\dagger \Phi(x)$$

(5.a) $$= (\partial\Phi/\partial t)^\dagger (\partial\Phi/\partial t) - (\nabla\Phi)^\dagger (\nabla\Phi) - m^2 \Phi^\dagger \Phi,$$

and the corresponding Hamiltonian is

(5.b) $$H = (\partial\Phi/\partial t)^\dagger \partial\Phi/\partial t + (\nabla\Phi)^\dagger \bullet \nabla\Phi) + m^2 \Phi^\dagger \Phi.$$

We shall be interested in minimizing energies. The kinetic terms are always positive definite so to minimize the energy associated with H we must take the fields to be constants in space-time.

This Lagrangian yields the field equation

(6.) $$(\partial_t^2 - \nabla^2 - m^2)\Phi(x) = 0,$$

where again I have simplified the notation by using (x) for (x, t).

The solutions of the field equations for individual particle and antiparticle states with momentum p have energies $\sqrt{p^2 + m^2}$, establishing the interpretation of m as the particle mass.

The Lagrangian is invariant under the "global" gauge transformation $\Phi \to \exp(i\Lambda)\Phi$ where Λ is a real number. It is not invariant under the space-time dependant or "local" gauge transformation $\Phi \to \exp(i\Lambda(x))\Phi$. Under an infinitesimal transformation a term is added to the Lagrangian of the form

(7.) $$\delta L = i\Lambda_{,\mu} \left(\partial_\mu \phi^\dagger \phi - \phi^\dagger \partial_\mu \phi\right) \equiv \Lambda_{,\mu} J_\mu.$$

If we think of Λ as a dynamical variable, another field, we have the Euler–Lagrange equation for it

(8.) $$\partial_\mu \delta L/\delta \Lambda_{,\mu} = \delta L/\delta \Lambda.$$

The term on the left is the divergence of the current generated by the local gauge transformation, while the term on the right vanishes hence the current generated this way is conserved. This is a variant of what is known as a Noether theorem.

We can consider, given that from the Noether theorem, $\partial_\mu J_\mu = \nabla \cdot J(x, t) + \partial/\partial t J_0(x, t) = 0$,

$$(9.) \qquad 0 = \int d^3 x \, \partial_\mu J_\mu$$

If nothing strange happens at the boundary in space[b] this reduces to

$$(10.) \qquad 0 = \partial/\partial t \int d^3 x J(x, t)_0 \equiv \dot{Q},$$

where $\dot{Q}$ is the spatial integral of the charge density $J_0(x, t)$.

We now want to quantize this theory. Thus we write,

$$(11.) \qquad \Phi(x) = \int d^3 k (b_k^\dagger \Phi_k^{(-)}(x) + a_k \Phi_k^{(+)}(x)).$$

Here the Φ's in the integral can be taken as a shorthand for the Fourier transform coefficients, say $\exp(ikx)$ with some suitable normalization. The a's and b's are annihilation operators and their conjugates are creation operators. The only non-vanishing commutators among the a's and b's are

$$(12.) \qquad [a_k, a_{k'}^\dagger] = \delta^3(k - k') = [b_k, b_{k'}^\dagger].$$

The momentum $\pi(x)$ conjugate to Φ is given by

$$\pi(x) = h \partial_t (\Phi^\dagger(x))$$
$$(13.) \qquad = -ih \int d^3 k \omega_k (b_k \Phi_k^{(-)*}(x) - a_k^\dagger \Phi_k^{(+)*}(x)).$$

This yields the commutation relation

$$(14.) \qquad [\Phi(x), \Pi(y)] = i\hbar \delta(x - y).$$

The charge can be written in terms of the a's and b's as

$$(15.) \qquad Q = \int d^3 p (a_p^\dagger a_p - b_p^\dagger b_p).$$

---

[b]For a discussion of the boundary question see "Spontaneous, Symmetry Breaking, Gauge Theories, the Higgs Mechanism and All That" by Jeremy Bernstein, *Rev. Mod. Phys.*, **46**, 1–48 (1974).

From this expression it follows that

(16.) $$[Q, \Phi] = \Phi,$$

so that

(17.) $$\exp(i\Lambda Q)\Phi \exp(-i\Lambda Q) = \exp(i\Lambda)\Phi.$$

Thus the charge generates the global gauge transformation, that is Φ transforms under unitary transformations generated by operator $\exp(i\Lambda Q)$.

The Hamiltonian in normal ordered form with the $a^\dagger$ and $b^\dagger$ to the left of the a and b is given by

(18.) $$H = \int d^3 k \omega_k (a_k^\dagger a_k + b_k^\dagger b_k).$$

We have dropped here an additive constant. Additive constants to the Hamiltonian do not change the physics since we are concerned with energy differences and not with absolute energies. (On the other hand an additive constant to the Lagrangian changes the action and thus the transition amplitudes which are computed with path integrals. We must make sure that this does not change the physics by eliminating this term.) The vacuum state $|0\rangle$ is that state for which the energy is minimum. If it has the property that $\langle 0|H|0\rangle = 0$. It follows that from such equations as $\langle 0|a_k^\dagger a_k|0\rangle = 0$ that

(19.) $$a_k|0\rangle = b_k|0\rangle = 0$$

so both the Hamitonian and the charge acting on this state is zero. We also have the obvious property

(20.) $$\langle 0|\Phi|0\rangle = 0.$$

So far the particle mass has simply been specified without any explanation for its origin. We now want to change scenes and introduce mass generation through spontaneous symmetry breaking. We introduce a new Lagrangian

(21.)
$$L = 1/2\partial_\mu \Phi(x)^\dagger \partial_\mu \Phi(x) + m^2/2\Phi(x)^\dagger \Phi(x)$$
$$- \lambda/4(\Phi^\dagger(x)\Phi(x))^2.$$

Several things are evident about this Lagrangian, Firstly, the term proportional to m^2 is not a mass term. Compare the sign to the sign in the mass term in Eq. (5a.). It is a self interaction term. Secondly, the Lagrangian is invariant under the global gauge transformations but not the local ones. There is a conserved current as before and a conserved charge. But does this charge annihilate the vacuum? i.e., $Q|0\rangle = 0$ as above, and if not, what does this mean? Here we run into the question of what is the vacuum?

We recall the equation

(22.) $$[Q, \Phi] = \Phi$$

which also holds here. What this implies is that if Q does not annihilate the vacuum then it must be that

(23.) $$\langle 0|\phi|0\rangle \neq 0.$$

This means that Φ cannot have a particle interpretation.

We recall from the discussion following Eq. (5a.) that the energy is minimized for constant fields and it is therefore determined by minimizing the potential. Let us consider the classical potential

(24.) $$V = -m^2/2(\Phi^\dagger\Phi) + \lambda/4(\Phi^\dagger\Phi)^2.$$

Clearly one extremal is when $\Phi = 0$. But quantum mechanically we want to replace this condition by the statement that the vacuum expectation value of the potential is a minimum. Indeed the vacuum is that state whose expectation value for the potential is a minimum. We shall see that in this case there is no unique answer.

Let us warm up with a simpler case which will illustrate the issues. Let us take a real field Φ so that the potential is

(25.) $$V = -m^2/2\Phi^2 + \lambda/4\Phi^4$$

The potential and the Φ's are all functions of the space time point x but at the minima of the vacuum expectation value of the energy they are constants so we can find a condition on the Φ's which minimizes the potential for all x. We take the derivative with respect to Φ and set it equal to zero. Thus

(26.) $$\Phi(m^2 - \lambda\Phi^2) = 0.$$

This has three solutions

(27.) $$\Phi = 0 \pm \sqrt{m^2/\lambda}.$$

This corresponds to the values of the potential of 0, which is a local maximum, and $\pm 1/4\,m^2/\lambda$, two distinct minima with the same energy. In the quantum mechanical case the vacuum is defined as that state in which the vacuum expectation value of the potential is minimized. In this instance there are two such states. If we pick the one with the positive minimum for Φ then for this vacuum

(28.) $$\langle 0|\Phi|0 \rangle = \sqrt{m^2/\lambda} \equiv v.$$

This means shows that Φ does not have the usual particle interpretation, and suggests that we introduce a new field η to describe the fluctuations of Φ away from its constant vacuum value v. Thus

(29.) $$\eta = \Phi - v.$$

In terms of this field the Lagrangian becomes

(30.) $$L = 1/2\partial_\mu\eta\partial_\mu\eta - m_\eta^2\eta^2 - \lambda v\eta^3 - 1/4\lambda\eta^4 + m^4/4\lambda$$

where

(31.) $$m_\eta = \sqrt{2m^2}.$$

We can make several remarks. This choice of vacuum has produced an η particle with a mass and some peculiar self interactions. But note that the $\Phi \rightarrow -\Phi$ symmetry of the original Lagrangian in Eq. (21) has been broken spontaneously. There is no trace of it in the transformed Lagrangian. The last term which is a constant also deserves further comment. If this was a Hamiltonian we could add a constant term with no misgivings. It would simply add a constant to all the energies. Since what we measure are energy differences there is no effect on the physics. But as I have mentioned the Lagrangian is a different matter. From it we define the action $\int_t^{t'} L dt$. If we add a constant to the Langrangian this adds a term proportional to the time difference in the action. We had better eliminate this term if we want a sensible theory.

With this example in mind we can now return to our complex fields with the continuous global gauge transformation invariance. As we shall see this brings in something new. The way to deal with this case is to write

(32.) $$\Phi = 1/\sqrt{2}(\Phi_1 + i\Phi_2)$$

where $\Phi_{1,2}$ are real fields. In terms of these fields the Lagrangian become

(33.) $$L = 1/2(\partial_\mu \Phi_1)^2 + 1/2(\partial_\mu \Phi_2)^2 + 1/2m^2(\Phi_1^2 + \Phi_2^2) \\ - 1/4\lambda(\Phi_1^2 + \Phi_2^2)^2.$$

Now the minima are given by the condition that

(34.) $$\Phi_1^2 + \Phi_2^2 = m^2/\lambda \equiv v^2.$$

The phase is undetermined. We can chose the phase for simplicity so that at the minimum

(35.) $$\Phi_1 = \sqrt{m^2/\lambda}, \quad \Phi_2 = 0.$$

We can then displace Φ_1 by its vacuum expectation value in the vacuum defined by this choice of phase and thus write

(36.) $$\Phi(x) = 1/\sqrt{2}(v + \eta(x) + i\xi(x)).$$

We can rewrite the Lagrangian in terms of these fields. There will be self-interaction terms of the η, ξ as well as interactions between them and the additive constant. What interests us is the "kinetic" term L_K.

(37.) $$L_k = 1/2(\partial_\mu \xi)^2 + 1/2(\partial_\mu \eta)^2 - m^2\eta^2$$

which shows that the new ξ field is massless while the η field has mass m.

Let us review what has happened. We began with a Lagrangian for a complex field of zero mass which was globally gauge invariant. We broke this gauge invariance spontaneously and ended up with two real scalar fields in interaction. One of these fields has mass zero and the other has acquired a mass. The question is, is this some freakish artifact of this Lagrangian or are we in the presence of a more general phenomenon? The answer is the latter. We have found a realization of what is known as the "Goldstone theorem."

I will not here try to give a detailed proof of this theorem but here is a statement of what it is. There are fine points which I will come to shortly. Suppose you have a theory with a certain number of conserved currents. Suppose these currents give rise to conserved charges that generate some set of gauge transformations. If one of these charges has a non-vanishing expectation value so that the gauge symmetry is broken spontaneously then necessarily this will give rise to a mass zero, spin zero particle — the ξ (xi) in the example above. On its face this would appear to rule out theories of this kind in elementary particle physics since there are no such particles. However, there is a loophole and through it we will drive a Mack truck. The loop hole is Lorentz invariance.

Needless to say we want all our theories to be Lorentz invariant but they need not be "manifestly" Lorentz invariant. The case in point is electrodynamics. This theory is certainly Lorentz invariant. Indeed when Einstein had to choose between Newtonian mechanics and electromagnetism he chose the latter precisely because it was relativistic. But electromagnetism is not "manifestly" Lorentz invariant in the following sense. The photon field A_μ is not well-defined. The theory is invariant under gauge transformations of the form $A_\mu \to A_\mu + \partial_\mu \Lambda$ where Λ is a function of the space time point x. This invariance precludes terms like $A_\mu A_\mu$ in the Lagrangian so the photon has no mass. To define the theory we must select a gauge. Two popular gauges are the "Lorenz gauge"[c] with $\partial_\mu A^\mu = 0$ and the "Coulomb gauge" with $\nabla \cdot A = 0$. The first gauge condition is manifestly Lorentz invariant while the second one is not. You can use either gauge to carry out computations. You will get the same answers for any physical quantity and these answers will be Lorentz covariant.

---

[c]This is not a misprint. Ludvig Valentin Lorenz was a 19th century Danish physicist to whom we owe this choice of gauge. I am grateful to Wolfgang Rindler for making this clear.

The proof that I know of that most clearly makes use of the manifest Lorentz covariance is due to Walter Gilbert.[d] Gilbert has an interesting history. He got his PhD in physics from Abdus Salam and then switched into biology. In 1980 he won the Nobel Prize for chemistry. It was during his physics period when he published this proof. I am not going to go into detail. An interested reader can look in the reference I gave in my 1974 review article.

The first people to use the gauge loophole were Peter Higgs[e] and Francois Englert and Robert Brout.[f] Many of the points were later clarified by G.S. Guralnik, C.R. Hagen and T.W.B. Kibble.[g] I will stay within the confines of the electrodynamics of charged scalar particles for the moment so as to use the work we have already done.

We can write the Lagrangian as

$$L = -1/4(\partial/\partial x_\mu A^\nu(x) - (\partial/\partial x_\nu A^\mu(x)))^2$$
$$- ((\partial/\partial x_\mu + ieA^\mu(x))\varphi^\dagger(x))(\partial/\partial x_\mu - ieA^\mu(x)\varphi(x))$$
$$(38.) \qquad + m^2\varphi^\dagger(x)\varphi(x) - 1/2(\varphi^\dagger(x)\varphi(x))^2$$

The first part of the Lagrangian is the free electromagnetic part and the last part is the bosonic part we have already seen. The middle part is the coupling term. As before is it convenient to split φ into its real and imaginary parts. Indeed we shall write in a two dimensional notation

$$(39.) \qquad\qquad \Phi = (\Phi_1, \Phi_2).$$

To economize the notation we shall introduce the 2×2 matrix q

$$(40.) \qquad\qquad q = \begin{pmatrix} 0 & -i \\ i & 0 \end{pmatrix}$$

---

[d]W. Gilbert, "Broken Symmetries and Massless Particles", *Phys. Rev. Lett.* **12**, 713–714 (1964).

[e]P.W. Higgs, "Broken Symmetries and the Masses of Gauge Bosons" *Phys. Rev. Lett.* **13**, 508–509 (1964).

[f]F. Englert and R. Brout, "Broken Symmetries and the Mass of Gauge Vector Mesons", *Phys. Rev. Lett.* **13**, 321–323, (1964).

[g]G.S. Guralnik *et al.* "Global Conservation Laws and Massless Particles", *Phys. Rev. Lett.* **13**, 585–587, (1964).

Using the Noether techniques I described earlier there is a conserved current

(41.) $J_\mu = i(\partial/\partial x_\mu \, \varphi(x) q \varphi(x) + e\varphi(x)' \varphi(x) A^\mu(x))$,

whose charges generate global gauge transformations. We can now break this invariance spontaneously and make the same choice of vacuums as before. This gives us the field equation, with the same definition of ξ as before.

$$\partial/\partial x^\nu (\partial/x_\mu A^\nu(x) - \partial/x_\nu A^\mu(x)) = m^2(1/m \partial/\partial x_\mu \xi(x) - A^\mu(x)).$$
(42.)

By this point the reader may wonder where is all this formalism leading? Behold, you are about to witness a miracle!

It is at this point we must choose a gauge for $A^\mu(x)$. It is convenient to take the Lorenz gauge $\partial/\partial x^\mu A^\mu(x) = 0$. Equation (43.) can then be written

(43.) $\partial/\partial x^\nu (\partial/\partial x^\nu A^\mu(x)) = m^2(A^\mu(x) - 1/m \partial/\partial x^\mu \xi(x))$

Let us define a new field $B^\mu(x)$

(44.) $B^\mu = A^\mu(x) - 1/m \partial/\partial x^\mu \xi(x))$

But this $\xi(x)$, unlike the previous example — Eq. (33.) et seq — obeys the equation of an uncoupled zero mass boson — the Goldstone;

(45.) $\partial/\partial x^\nu (\partial/\partial x^\nu \xi(x)) = 0$

Hence Eq. (44.) becomes

(46.) $\partial/\partial x^\nu \partial/\partial x^\nu B^\mu = m^2 B^\mu.$

The "photon" has morphed into a vector meson with mass!

Let us summarize what has happened. The electrodynamics of a charged boson with a spontaneously broken gauge symmetry has in the manifestly covariant Lorenz gauge produced

1. a Goldstone theorem
2. an uncoupled massless Goldstone scalar boson ξ
3. a massive scalar boson η
4. a massive vector meson B_μ

Since these masses have the same origin there is a relationship between them. What happens in the Coulomb gauge div $\mathbf{A} = 0$? I won't go though the steps but summarize the results.

1. no Goldstone theorem since the gauge is not explicitly covariant
2. no Goldstone boson
3. a massive scalar boson η
4. a massive vector meson B_μ

The first person to make full use of these ideas was Steve Weinberg in 1967.[h]

To appreciate what he did we must set the context with some history. In 1934 Fermi produced the first modern theory of β-decay. He was an expert in quantum electrodynamics so it was natural for him to use it as a template for his new theory. In quantum electrodynamics the current of charged particles J^μ interacts with the electric field A^μ with a coupling of the form $J^\mu A^\mu$. Thus charged currents do not act directly with each other but only by the exchange of photons. On the other hand, since there was no equivalent of the photon for the weak interactions like β-decay, Fermi decided to couple a current for the "nucleons" — neutron and proton — J^μ_{N-}, with a current for the "leptons" — electron and neutrino — J^μ_L-directly, i.e., J^μ_N J^μ_L. As a phenomenological theory this worked very well. One could use it to compute, for example, the energy spectrum of the electrons emitted in β-decay. But it came to seem anomalous. The "strong" interaction between nucleons, as Yukawa proposed in those pre-quark days, took place with the exchange of mesons, the electromagnetic interactions with photons and presumably gravitation with gravitons. Indeed there was a suggestion of using the same meson that produced the strong interactions to produce the weak ones. This was abandoned. But in the 1950's it was suggested that one or more weak heavy photons might do the trick. There were two problems. None had been observed and the theory that was being proposed did not make any sense.

---

[h]S. Weinberg, "A Model of Leptons", *Phys. Rev. Lett.* **19**, 1264–1266 (1967).

The former difficulty was easily disposed of. Since the contact theory with the currents coupled directly to each other worked well phenomenologicaly it had to be that these weak mesons were very massive — too massive, it was argued for the generation of accelerators that then existed to produce them. Indeed, when they finally were produced it turned out that their masses were about a hundred times that of the nucleon masses. The second difficulty was a different matter. In the theories that were then being proposed the weak mesons were being put in "by hand." They were just massive particles whose masses had no particular origin. If one tried to compute anything beyond the lowest order phenomenology one got terrible infinities. These infinities were much worse than those in quantum electrodynamics which could be swept under the rug by renormalization. In short the theory did not make any sense. Theorists were left grasping for straws. Then came Weinberg.

The "electroweak" theory of Weinberg plays on the themes we have discussed but at a higher register. The underlying Lagrangian consists of massless vector mesons three of which — the two charged ones and the neutral one — are coupled to the scalar bosons. In addition there is the photon which is not coupled to these bosons. Then there are the bosons themselves which are self-coupled as well. This Lagrangian has a global gauge symmetry but the symmetry group is non-Abelian. Recall in the examples I gave before the effect of the global gauge transformation is to multiply the fields by a numerical phase. These example are Abelian so it does not matter in which order two of these transformations are performed successively. In the non-Abelian case it does matter. All of this complicates the formalism but does not change the underlying methodology. Once again the gauge symmetry is broken spontaneously. The coupled vector mesons acquire masses while the photon remains massless and there are massive scalar bosons. Apart from the fact that this unifies two otherwise disparate interactions it also cures the non-renormalization issue. There was always a sort of canary in the mineshaft interaction that was used as an illustration, it was the reaction $\nu + \bar{\nu} \rightarrow W^+ + W^-$. A neutrino and an anti-neutrino collide and produce a pair of weak vector mesons. No one proposed to measure this reaction but its calculation should nonetheless make sense.

When this calculation is done in the conventional theory with no scalar bosons and the masses being put in by hand the cross section increases without limit as the neutrino momentum tends to infinity. Here there are no issues of infinities caused by going to higher orders but rather of a violation on the limits that quantum mechanics imposes on the magnitude of such cross sections. What Weinberg observed was that in the electroweak theory there was a contribution from the scalar bosons to this process that cancelled against this and rendered the cross sections sensible. He conjectured that the theory was renormalizable and this was proven in detail by Martinus Veltman and his student Gerhard t'Hooft.

The alert reader will have noticed that something is missing in this discussion. All the leptons have mass including the neutrinos. To say nothing of the masses of the neutrons and protons. What is the origin of them? Perhaps the reader will indulge me in a bit of personal reminiscence. For two years in the late 1950's I was a postdoctoral at Harvard. Julian Schwinger was the leading light in theoretical physics at the time. We, the post docs and junior faculty, audited whatever course he happened to be teaching. The material was always original and new. The lectures were on Wednesdays and afterwards the small group of us would have lunch with Schwinger at *Chez Dreyfus* in Cambridge. We would be joined by another small group from MIT that included Vikki Weisskopf. If Schwinger had any new ideas he would try them out on Wiesskopf. As it happened on this occasion he had developed a "theory of everything." Some of this survives in the work of other people. For example in 1962 he published a paper called "Gauge Invariance and Mass."[i] In it he raised the question of one could have massive vector meson in a theory which had an underlying gauge invariance. This is not exactly what we have been discussing, but it did inspire P.W. Anderson to use these ideas in condensed matter physics.[j] Anderson uses language in a non-relativistic context which is very similar to what we have been discussing. I remember a particular lunch in which Schwinger

---

[i] J. Schwinger, "Gauge Invariance and Mass", *Phys. Rev.* **125**, 397–398 (1962).

[j] P.W. Anderson, "Plasmons, Gauge Invariance and Mass", *Phys. Rev.* **130**, 439–442 (1963).

began by saying to Weisskopf, "Now I will make you a world." The "world" was written down on a few paper napkins one of which I saved. In any event one of the things that he said which has stuck with me ever since was that scalar particles were the only ones that could have non-vanishing vacuum expectation values. He then went on to say that if you couple one of these to a fermion Ψ by a coupling of the form $\Phi\Psi^\dagger\Psi$, then this vacuum expectation value would act like a fermion mass. This is how it's done in principle. All particles in this picture would acquire their masses from the vacuum. We are a long way from Newton.

I have avoided so far the use of the term "Higgs boson" — the analogue in the electroweak theory of the η above. Certainly Peter Higgs deserves the credit for first exhibiting the mechanism in the context of scalar electrodynamics. But as I have tried to show it took other people to make it work. But the "Higgs boson" is what is being looked for at CERN. If they find it we shall all be happy and relieved. And if not? I am reminded of a story about Einstein. He had just received a telegram with the news that the eclipse expeditions had confirmed his general relativity prediction about the Sun bending starlight. He was very pleased with himself and showed the telegram to one of his students, Ilse Rosenthal–Schneider. She asked him what he would have done if the telegram had contained the news that the experiments disagreed with the theory. He replied, *"Da könt mir halt der lieber Gott leid tun, die theorie stimmt doch."* — "Then I would have been sorry for the dear Lord. The theory is right."

18. An Upper Bound

In the previous essay I presented the theory of the so-called Higgs mechanism for generating mass, which implies the existence the Higgs boson. What I did not point out was that under generally accepted assumptions this boson has an upper bound to its mass. This very remarkable prediction was the basis for believing that it might be discovered in experiments using existing accelerators running experiments. The purpose of this letter is to give a primitive explanation of why this is true. Before doing so let me note that Pauli in his famous 1930 letter to the "Radioactive Ladies and Gentleman" predicted that there would be an upper bound to the mass of what he called the "neutron," later re-named by Fermi the "neutrino." Pauli's bound was 0.01 proton masses and was derived from a consideration of the conservation of energy in beta decay. The bound here is much more theory dependant.

In the Higgs Lagrangian there is a term $m^2\varphi^\dagger\varphi - 1/2f^2(\varphi^\dagger\varphi)^2$. Here f is a dimensionless coupling constant and φ the Higgs field. Note that because the sign of the first term is positive, m is not a mass. The key to the Higgs mechanism is that the vacuum expectation value of the Higgs field is not zero. Indeed let us call this vacuum expectation value η. It has the dimension of a mass and indeed all the masses in the theory are manufactured from the vacuum energy. This field does not have a particle interpretation but the shifted field $\varphi' = \varphi - \eta$ does. Its mass ignoring all radiative corrections is given by, $m = \sqrt{2}\eta$ which follows from the shifted Higgs Lagrangian. This humble formula, which says that the mass is proportional to the dimensionless coupling constant, is the key to everything.

To obtain an upper bound on the mass two assumptions are made: The Standard Model is viable and that its predictions can be evaluated in perturbation theory. The second assumption requires that $f < 1$ which means that $m < \sqrt{2}\eta$. To make use of this we must know η. Here is where the Standard Model comes in. The same vacuum expectation value gives rise to the W mass so that again ignoring all radiative corrections one finds $m_W = g\eta\sqrt{2}$, where g is the dimensionless weak coupling constant. Because we know g (about 0.65) and m_W (about 80 GeV), we can estimate η, and find the upper bound for the Higgs mass, which turns out to be about 240 GeV. This very crude calculation has been improved on by many people, and the consensus upper bound is about 285 GeV. But I think the crude calculation makes it clear why there is an upper bound. See the next essay for the experimental result.

19. A Higgs Memorandum

In the previous essay I derived an upper bound to the Higgs boson mass. This was an application of the Standard Model. In deriving this bound, the assumption was made that perturbation theory was valid and that the coupling constant of the Higgs was less than one. In this model, the coupling constant is proportional to the mass, which explains the bound that turned out to be about 240 GeV. This was a very crude calculation. In the summer of 2012 a particle was identified that has a mass of about 126 GeV and appears to have at least some of the properties of the Higgs boson. The purpose of this essay is to describe what is presently known experimentally and to signal what further experiments are necessary.

There is nothing original in what I am going to say, but it may be of some pedagogical service.

The particle appears as a resonance in the proton–proton collisions produced in the Large Hadron Collider at CERN. The two protons have a total charge two but the Higgs is electrically neutral so the first question is what has happened to the charge? At the extremely high energies of these collisions, the protons appear to each other as quarks and gluons. In some of the collisions two gluons, which are electrically neutral, can fuse into quarks that in turn produce the Higgs. The positive charge is carried off in the detritus of quarks that are detected as background. The particle produced has a very short lifetime and several decay modes, which hopefully

271

will reveal its identity. I am going to confine myself to just the decay mode into two gammas and the decay modes into four leptons.

I begin with the decay into two gammas and always work in the rest frame of the decaying particle. The experimenters know enough about the kinematics of these decays to transform their results to this frame. In the Standard Model, the spin of the Higgs is zero and its parity is positive; this is what the experiments should confirm. The first remark is that the fact that it decays into two photons shows that the spin cannot be one and also that the newly discovered particle is electrically neutral. The spin-one prohibition is what is known as "Yang's theorem" and is simply proved by trying to construct the two-photon spin-one state. There are two gamma polarization vectors (e_1 and e_2) that are transverse to the photon momentum k and there is the Higgs polarization vector e_H that is not transverse to k. Thus, there are three possible terms in the construction of the spin one state: $e_1 \times e_2 \cdot e_H$, $e_1 \times e_2 \cdot k e_H \cdot k$, and $e_1 \cdot e_2 e_H \cdot k$. None of these are symmetric under the exchange of the two gammas demanded by the quantum theory, which establishes the result. Spin two is possible. If S_{ij} is the symmetric spin tensor then you can have a term like $S_{ij} e_{1i} e_{2j}$. There is no theoretical imperative for a spin-two Higgs, but it is not yet entirely ruled out experimentally. This leaves us with spin zero.

Let us continue with the two-photon mode. There are two ways we can compose the final state, remembering that the polarizations are transverse to the momentum: $e_1 \cdot e_2$, and $e_1 \times e_2 \cdot k$. The first is a scalar and the second is a pseudoscalar. The same issue comes up with the two-photon decay of the neutral pi-meson. For these very-high energy photons, the polarizations are not readily measurable directly. But the photons can convert into electron–positron pairs that are known as "Dalitz pairs." The momentum vectors of the particles in each pair are in a plane and the azimuthal angular correlations of these planes depend on the parity. This is one way the parity of the neutral pion was determined. In principle one could do the same thing with the Higgs but the double-Dalitz pairs are a small correction to the two gamma decay so in practice the effect may not be observable and one must resort to different methods. By the way, one may wonder how a scalar and a pseudoscalar boson can both decay into two gammas if parity is conserved in the decay. If R is the right circular polarized photon and L the left circular one, then the scalar state is RL + LR while the pseudoscalar state is RL − LR.

The Higgs can decay into two oppositely charged W mesons or two neutral Z mesons. The W mass is about 80 GeV and the Z mass is about 91 GeV, meaning that a pair of either is more massive than the Higgs. Hence one or both of these mesons in the decay must be "virtual," which means that instead of it being produced at its rest mass value, there will be a spectrum of virtual Z meson masses produced in these decays. I am going to focus on the Z decays since the electron–positron or the muon–antimuon pairs that are produced are all directly detectable so that the kinematics can be nailed down accurately. The W decays produce neutrinos so they are more complicated to nail down. In these Z decays a spectrum of virtual masses can be determined and this spectrum for reasons I will explain is sensitive to the parity. But first I want to give a theoretical argument that suggests that the observed particle is a scalar.

Suppose the particle is a pseudoscalar and consider how it can enter the Lagrangian. To set the stage, I remind you of something from electromagnetism. If A_μ is the electric field, then we can define the field tensor $F_{\mu\nu}$ as $\partial\mu A\nu - \partial\nu A\mu$, where the derivatives are taken in the various space–time directions. This is an antisymmetric tensor and its dual can be defined as $\tilde{F}_{\alpha\beta} = \varepsilon_{\alpha\beta\mu\nu}F_{\mu\nu}$, where $\varepsilon_{\alpha\beta\mu\nu}$ is the totally antisymmetric Levi-Civita symbol and it is assumed that we sum over repeated indices. The product $\tilde{F}_{\mu\nu}F_{\mu\nu}$ is a pseudoscalar. In the electromagnetic case, it contains terms like $E \cdot B$. Here, we can repeat these steps with the field Z_μ. We can couple the product of the field tensor and its dual to the putative pseudoscalar Higgs field. However, if we put this coupling into the Lagrangian, the resulting field theory is singular. There are infinities that cannot be renormalized away. Terms like this one can arise as higher-order corrections to a theory in which there is a scalar Higgs that couples directly to the Z's, which is what the Standard Model proposes. Because the observed four-lepton production seems to be consistent with the Standard Model calculation, this gives at least an indirect argument that the observed particle is a scalar. But one would like to see a direct experimental confirmation. Perhaps correlations of the type used to measure the parity of the neutral pion will become available as the data improve. The idea would be to study the planes formed by the momenta of the lepton pairs produced by the Z decays. The decays can be into two electron–positron pairs, two muon–antimuon pairs, or one of each. But these decays can be used in another way to measure the parity. As I have mentioned, because of their masses, one or both of the Z's must be virtual. By measuring the kinematics in each four-lepton decay, one can find and plot the frequency of these virtual masses as they occur. These plots will differ for scalar and pseudoscalar Higgs. Effectively the Lagrangian will contain couplings like the one I described above when I noted that the pseudoscalar coupling could not be a part of the fundamental Lagrangian, although terms like this can arise in calculations. The pseudoscalar produces higher powers of the Z-momentum than the scalar in the coupling. If we look at the kinematics, we readily see that the higher the momentum, the smaller the mass of the virtual Z's. Hence the pseudoscalar distribution is weighted toward smaller virtual masses. What one can say is that the present data are consistent with a scalar spin-zero Higgs. From a theoretical point of view, there is a lot at stake here. But one must always remember that at its base, physics is an experimental science and sometimes nature has a way of surprising us. There is something else to keep in mind. The study of self coupled Higgs-like bosons goes back a long way. There has been for many years the suggestion that these theories are "trivial." What this means is that they are only consistent when the coupling constants vanish. The coupling constants are a function of energy and if they do not vanish identically then it is argued that in these theories at very high energy they turn negative which renders the theory unstable. One might expect that manifestation of this would show up at the energies of the Large Hadron Collider which would mean that physics beyond the Standard Model had manifested itself.

Final Acknowledgements

The following essays were published in substantial part in *The American Journal of Physics*. They are copyrighted and reprinted with permission by the American Association of Physics Teachers in the years given below.

"Botherhood of the Bomb", **71**, 411 (2003)
"Bachelier", **73**, 395 (2005)
"Max Born and the Quantum Theory", **73**, 999 (2005)
"P.A.M. Dirac, Some strangeness in the proportion" **77**, 1175 (2009)
"A question of mass" **79**, 25 (2011)
"An Upper Bound", **79**, 810 (2011)
"A memorandum that changed the world" **79**, 440 (2011)
"More about Bohm's quantum", **79**, 601 (2011)
"A Higgs Memo", **81**, 5 (2013)

Two of the essays were first published in *Physics in Perspective: John von Neumann and Klaus Fuchs: an Unlikely Collaboration* — Jan 1, 2009 — and *The Drawing or Why History is Not Mathematics* — Jan 1, 2003. These are reprinted with the permission of Springer.

The photograph of Gernot Zippe is reprinted with the permission of Houston Wood. I would like to thank Carey Sublette for permission to use his annotated version of the drawing of the Fuchs von Neumann device on page 176.

I would also like to thank Jessica Barrows for her encouragement and G.H. Chong for his skilful editing to keep things simple.

Printed in the United States
By Bookmasters

Printed in the United States
By Bookmasters